AUTOMATION

Automation

the technology and society

RAPHAEL KAPLINSKY

Longman

Longman Group Limited,
6th Floor, Westgate House, Harlow, Essex CM20 1NE, UK

First published 1984

British Library Cataloguing in Publication Data

Kaplinsky, Raphael
Automation: the technology and society.
1. Computers and civilization
I. Title
306'.46 QA76.9.L66
ISBN 0-582-90203-7

Printed in Great Britain by
Butler & Tanner Ltd, Frome and London

For
CATHY

who always kept my mind on the
really important issues when I
was overawed by the technology

Contents

List of Tables

Figures

Preface

The issue of automation has been a matter of continuing debate throughout the post-war period. But in spite of this, as many authors have pointed out, the discussion has remained a very ill-structured and disorganised debate. In Chapter 2 Raphael Kaplinsky indicates some of the reasons for the lack of clarity in the debate, including the lack of agreement about definitions. One of the great merits of his book is that it provides a systematic and coherent framework within which to discuss the subject.

In his earlier work on computer-aided design Kaplinsky already made a substantial contribution to our understanding of automation. He has now extended this to cover three main spheres: design, manufacturing and coordination (Chapters 3 to 5). His distinction between 'intra-sphere' and 'inter-sphere' types of automation in these three sectors is particularly valuable, and enables the reader to grasp the full significance of inter-sphere automation, as well as some of the reasons for the apparently slow take-off of full-scale totally automated systems in the enterprise as a whole. His discussion in Chapter 6 of the technological, economic and social barriers hindering the introduction of this final stage of integrated automation seems realistic, although perhaps still somewhat underestimating the problems of divisive group interests and of management subcultures even in the more successful enterprises of Japan and Western Germany.

Another great merit of Kaplinsky's book is his resolutely historical approach. This helps a great deal to resolve some of the paradoxes, which are otherwise very hard to understand. It has enormous advantages over much of the technical and systems literature which abstracts completely from this historical context of automation. Without a sense of history, the diffusion of automation technology usually appears unaccountably slow, since its technical advantages have been clearly apparent to the professionals in the business for three decades.

The early pioneers of computerized automation, such as Wiener and Diebold, envisaged most of the contemporary developments in the 'automated factory', the 'electronic office', computer-aided design and management systems, as well as 'intra-sphere' automation. However they greatly overrated the speed of this vast social transformation, and of its employment effects.

They failed to take adequate account of the time-scale of investment and training required to build up an entirely new electronic capital goods industry, a huge components industry and a software supply system. They failed also to have sufficient regard for the problem of comparative costs and profitability of automation in relation to alternative investments. Even today, when the microprocessor revolution has vastly reduced the costs of automation in many areas of potential application, and when there has been an enormous expansion of the education and training programmes for computer scientists, technologists, technicians and programmes, there are still major bottlenecks and delays associated with shortage of skilled people and of capital.

Even now, as Kaplinsky shows, the full-scale automation of most manufacturing and service activities has only just begun. By his full recognition of the long-term factors involved in the diffusion of a major new technological system, he avoids some of the pitfalls which have attended much of the discussion on the employment effects of automation. The fears of mass unemployment associated with automation, which were widespread in the 1950s and 1960s, proved to be ill-founded at that time, but as Kaplinsky points out in Chapter 1 and in his more detailed discussion in Chapter 8, the economic context in which automation is now diffusing is very different from the post-war boom. Paul Einzig, who in his book on *The Economic Consequences of Automation* in 1957 justifiably criticized many of the simplistic arguments about automation leading rapidly and inexorably to mass unemployment, nevertheless commented:

> Although fears that automation is liable to lead to slump through over-investment are unfounded, there can be no doubt that, if deflation should develop through no matter what causes, automation would greatly exaggerate any downward trend in purchasing power, prices and employment. Just as under inflationary conditions it strengthens the influences making for a boom, under deflationary conditions it would be used for reducing the number of employees. . . . Once the slump is followed by chronic depression, automation would resume its course, almost entirely with the object of reducing costs of production. (p117–118).

Deep depressions, such as those of the 1930s and 1980s, may be seen as periods when the mismatch between social institutions and the potential of revolutionary new technologies is particularly evident. There is no deterministic solution to these problems today, any more than there was in the 1930s. But whatever political and social solutions are attempted in the 1980s and 1990s they will have to take into account both the technological and social issues identified by Kaplinsky. It would in principle be quite possible to use automation technologies to reinforce the tendencies towards deskilling of the labour force, sophisticated supervision by 'Big Brother' techniques of every second of the working day and a high degree of authoritarian centralization. But it would also be possible to achieve the opposite solution of 'automation with a human face'.

Full recognition of the social, economic and behavioural factors affecting automation is one of the greatest strengths of Kaplinsky's analysis, whether with respect to overall employment effects, skill effects, or the international division of labour. It is a book which is essential reading for anyone wishing to understand the interplay of social and technological factors. It is especially important for its recognition of the element of social choice in shaping the new technologies and mode of application.

The book does not discuss the automation of distribution, warehousing and transport. Nor does it discuss the automation of tertiary activities, such as financial services and social services. I very much hope that he will follow his first two books with others which take up these questions. But meanwhile as an analysis of contemporary trends in the automation of manufacturing enterprises this is a first-rate original contribution.

Professor C. Freeman
Science Policy Research Unit
University of Sussex

Acknowledgements

We are grateful to the following for permission to reproduce copyright material:

Association of Professional, Executive, Clerical & Computer Staff for fig 2.7 from p. 388 'A Trade Union Strategy of the New Technology' *Apex* 1979; Audits of Great Britain Ltd for Table 1.3 from *Home Audit Facts and Figures* 1980; Benn Electronics Publications Ltd for fig 2.10 from 'A prognosis of the impending intercontinental LSI battle' I. M. Mackintosh, *Microelectronics Journal* Volume 9, No 2, Dec 1978; David & Charles Ltd for table 1.4 from *The World Automative Industry* by G. Bloomfield; The Financial Times for the article 'Some Workers have already accepted wage cuts' *Financial Times* 15.9.82, extract from the article by J. Griffiths 'FIAT: Now Robots for Engine Assembly' *Financial Times* 2.11.81, Tables 4.4, 4.6 from Table p. III 'Supplement on Manufacturing Automation' *Financial Times* 16.7.82; W. H. Freeman & Co for figs 2.8, 2.9 from pp 67 and 69 of Robert N. Noyce 'Microelectronics' *Scientific American* 1977; Copyright © 1977 by Scientific American Inc. All rights reserved; the author Dr J. Gershuny for fig. 8.2 and an extract from p. 1 *Technical Change and Sectoral Development* 1979; Harvard University, Graduate School of Business Administration, for fig 2.2 from p. 45 Ex 4–2 and extracts from pp. 20, 118, 223 from James R. Bright *Automation and Management* 1958; Institute of Manpower Studies for Table 7.6; the author Stanley Klein for an extract from *The S Klein Newsletter on Computer Graphics* Vol. 4 No. 6 22.3.82; Macmillan Accounts & Administration Ltd for Table 7.1 from Table 7 'Variations in Production costs in relation to Different Scales etc.' from *Technology and Underdevelopment* by Francis Stewart; Organisation for Economic Co-operation and Development for Tables 1.6, 1.7 and an extract from *Technical Change and Economic Policy: Science and Technology in the New Economic and Social Context* 1980, Table 1.5 from Table 14 p. 34 *Interfutures Draft* Final Report Part V Fut (78) 10, 1979, and fig 1.1 from Diagram 6.2 by P. Hill *Profits and Rates of Return* 1979; Frances Pinter Ltd for Tables 1.1 and 1.2 from Tables 1.1a, 8.1 pp. 2, 148 *Unemployment and Technical Innovation: Study of Long Waves and Economic Development* by Freeman et al 1982; Science Policy Research Unit, Sussex University Research Centre, Brighton for Table 4.2 of *The Metalworking Machine Tool Industry in Western Europe and Government Intervention* by D. T. Jones 1980; Sussex University Press/IETB for Table 4.1 and extracts p. 58 *Changing Technology and Manpower Requirements in the Engineering Industry* by R. M. Bell, 1972; World Bank for Table 9.2 from Table 2.4 p. 12 in *World Development Report* 1982 pub OUP Inc.

We have been unable to trace the copyright owners of an extract and figure of G. Jacobs 'Designing for Improved value' *Engineer*; and fig 5.2 Table 7.7 from *Japan Foreign Press Center* R–82–10 October 1982 and would appreciate any information that would enable us to do so.

Introduction

This book has a number of objectives and aims to operate at a number of levels, combining detailed analysis with broader overview, and technical discussion with social comment. It is divided into three parts. In the first part the historical link between the diffusion of automation technologies and the progress of economies is discussed. Whilst it is true that our understanding of the concept of automation is in some senses in a state of crisis (given the widely varying interpretations of the term) the 'crisis' referred to in the first chapter of this book is of a different form. What I have tried to show is that the often cited effect that the diffusion of automation technology creates economic crisis by displacing labour turns real events on their head. What is in fact happening is that the emergence of economic recession and depression in the late 1970s and early 1980s has occurred for autonomous reasons, having to do with the internal workings of our system of economic organization. The consequence has been a sharp rise in competitive pressures and a decline in profit rates. A large number of firms are responding to these circumstances by adopting the new automation technologies which are based upon maturing electronics technologies. Thus automation is considered as arising out of economic crisis, rather than causing it.

Part Two is more technical in content. It is based upon a review of the automation literature which has hitherto largely concerned itself with automation within manufacturing proper. Almost no attempt has been made to link this discussion with emerging automation technologies in design and in the office. Not only is this a serious omission in terms of coverage of technological trends, but it obscures what is probably likely to be the most significant technological-cum-organizational development of the last two decades of the twentieth century. Namely, when digital-logic control systems in individual activities in each of the three spheres of production (that is manufacture, design and coordination) are linked together, the primary benefit realized by successful innovators is systems-gains. It is this fact which is forcing the restructuring of major transnational corporations such as General Electric, Westinghouse, IBM and AT and T and leading to the development of the 'factory of the future'.

In Part Three we assess the 'impact' of the new automation technologies on society. We see how its introduction is associated with an increasing concentration of capital, a destruction of craft skills and displacement of labour, and a whittling away of Third World comparative advantage based on low-labour costs. All this leads to the question 'Is the technology to blame?'. Thus in the concluding chapter we consider whether these negative impacts have in some sense been inevitable, or whether they have reflected the types of social organization within which they have been introduced. If it is the former then our social concern must surely lie in limiting the diffusion of the new automation technologies. But if it is the latter then our concern is better placed in challenging the social environment in which the technology is being developed and diffused.

In convering such a wide territory I have inevitably been forced to consider a broad range of literature, drawing from economics, engineering, sociology, management theory and Marxian political-economy. The disadvantage of this is that in many cases readers are confronted with a 'strange' set of literature which seems too technical for their specific needs or interests. However, I hope that this is outweighed by the fact that readers will be exposed to ideas that do not normally confront them. Thus, for example, engineers will be forced to question whether the particular techniques they are involved with are inevitable, or for the public 'good'; economists will realize that technical progress is not 'manna' from heaven but requires intensive firm-level development; and that sociologists will be exposed to the relationship between social relations and technology, rather than merely focusing upon the social impact of technology. The overwhelming influence of the military imperative in the development of automation technologies over the years continues to send a chill down my spine; I hope I have conveyed this concern to my readers.

In writing this book I have benefited from the help of many people. My primary intellectual debt is owed to Chris Freeman whose work on the importance of electronics has stimulated me constantly over the past five years. Ian Miles and Hubert Schmitz were particularly helpful in their detailed commentary on earlier drafts. I also benefited from comments from John Bessant, Brian Easlea, Tom Husband, Kurt Hoffman and John Irvine and from the many people in industry who have helped me over the years and given me access to their plants and time. I am also grateful to Karen Brewer and Karen Sage who willingly typed and retyped the various drafts. Finally my primary thanks are due, as always, to my family for continuing to put up with my drafting difficulties, and in particular to my wife, to whom I have dedicated this book.

PART ONE

Understanding Automation

CHAPTER ONE

Crisis and Automation

What makes one of the world's five largest industrial companies change its corporate strategy, moving away from an historic emphasis on consumer products, nuclear power plants and jet engines to provide the 'factory of the future' for American and European industry?

Consider the case of General Electric (GE), the largest engineering company in the world. With its subsidiaries producing a wide spectrum of products, ranging from coal to domestic refrigerators, jet engines, nuclear power plant and numerical controls, GE had a turnover of over $27 billion in 1981 and employed over 400,000 people worldwide. Its profits in that year were $1.5 billion, despite the expenditure of over $1 billion on research and development. Yet this appearance of success was not enough to satisfy the new management which took over in spring 1981. Careful examination of the company's performance had revealed three major problems which seemed likely to confront GE in the 1980s. First, nuclear power plant orders were declining for a variety of reasons (for example declining growth rates for electricity demand, increasing environmental concern and technical problems), and what had earlier seemed to be a major sector of potential growth now looked more uncertain. Second, failure to keep pace with new technology in the traditional consumer products division – notably in the use of electronic controls to be incorporated in the final product, and in the installation of automated plant to cut costs and improve product quality – suggested that unless something was done quickly, GE would soon lose its dominance of the US market. The fact that the consumer products division contributed 22 per cent of GE's total profits, whilst only accounting for around 16 per cent of total sales, lent urgency to the need to upgrade the division's competitiveness. And third, even in the higher technology divisions, GE had displayed a tendency to rest on its technological laurels and was feeling the effects in final markets. For example, in the US machine tool numerical controls market, traditionally dominated by GE, a Japanese-German joint venture captured 15 per cent of the market in its first five years and aimed for a 25 per cent share within the next five. These advances were almost entirely due to its prowess in taking advantage of the potential offered by advanced electronics-based technologies.

Faced with this bleak outlook on the future, GE has instituted a major change in corporate strategy. Rallying behind the call 'automate, emigrate or evaporate', GE is aiming at the 'factory of the future'. In part this involves the installation of automated production plant in its own manufacturing operations, but more significantly GE is aiming to become the major US supplier of automated 'factories of the future' which it believes will need to be installed in the 1980s if the US economy is to survive competitive pressures. Amongst other steps, this strategy has involved the following key acquisitions:

- purchasing the third largest computer-aided design supplier (Calma) for $170 million in 1981, ten times the price paid two years earlier by its owners;
- acquiring a major manufacturer of metal-oxide integrated circuits, Intersil, for $235 million; these particular types of components are especially well-suited to harsh factory environments, and hence for automation equipment. (In the early 1970s GE had actually disposed of its electronic subsidy);
- developing its own manufacture of industrial robots initially through technology licences from foreign firms such as Volkswagen and Hitachi; GE is aiming for 20 per cent of the US robot market by 1986 and for 30 per cent by 1990;
- beefing up its sagging numerical controls and industrial controls divisions by investing over $100 million and establishing a Microelectronics Research Centre at an additional cost of over $50 million;
- through an acquisition of 48 per cent of the equity in Structural Dynamics Research Corp (SDRC), GE will establish a series of productivity centres throughout North America and Europe. These will offer the combined service of GE subsidiaries in helping downstream user-industries to capitalize on the potential offered by new automated equipment;
- acquiring a number of smaller computer service bureaux for over $100 million, to make GE's own subsidiary GEISCO, the largest computer service firm in the world;
- selling off subsidiaries in other areas. The Australian coalmining offshoot was sold for $2.4 billion and the air-conditioning subsidiary for $135 million.

This radical change in structure in which GE's 'strategy is to become the number one integrator of factory automation and the number one solution producer' represents an extreme reaction to the changing economic climate of recent years, and cost the company upward of $700 million in acquisitions and $2 billion in automating its own plants. But it is not unique and many of GE's competitors, in the US as well as Europe and Japan, are making similar decisions. Westinghouse, for example, another large US engineering firm, describes its new strategy as covering 'the entire spectrum of factory automation and includes: processing information through computer-aided design, computer-aided manufacturing and computer-aided testing; productive machinery including robots, machine tools and material handling equipment; and the communication links that connect these islands of automation.' IBM, too, with its recent introduction of microcomputers and industrial robots (made actually to Japanese designs) clearly shares a similar perception of future strategy.

But why this major change in strategy, concentrated as it is into such a short

Table 1.1 Average annual growth rates of gross domestic product

	1870–1913	*1913–50*	*1950–60*	*1960–70*	*1970–80*	*1973–80*
France	1.7	1.0	4.7	5.6	3.5	2.8
Germany	2.8	1.3	8.1	4.8	2.8	2.4
Italy	1.5	1.4	5.1	5.3	3.1	2.8
Japan	2.5	1.8	8.6	0.3	4.7	3.2
UK	1.9	1.3	2.7	2.7	1.8	1.0
USA	4.1	2.8	3.2	4.2	2.9	2.1

Source: Freeman, Clark and Soete (1982)

period of time? In order to understand this transition and see its full significance, it is important to delve back briefly into post-war economic history. As we can see from Table 1.1, the era of post-war economic expansion, was historically unprecedented, particularly in the 1950–1970 period. At the same time most of the world's developed economies experienced levels of near-full employment (Table 1.2). Again, this was historically unusual, especially compared to the era between the first and second world wars, when in Britain, for example, in only one year did unemployment fall below 10 per cent of the labour force. The unprecedented expansion of the world economy was most evident in relation to trade: whereas total global production in manufacturing increased by 317 per cent between 1953 and 1977, global trade in manufactures increased by 673 per cent in the same period.

However, around 1970–75, 'things began to go wrong'. The rate of growth of almost all the major economies began to decline (see Table 1.1); unemployment rose to near-1930s levels (see Table 1.2); industrial output actually declined in many economies particularly in the US and the UK in the early 1980s; and for only the second time in the post-war period the volume of world trade actually fell in 1981. (It also fell in the recession in the mid 1970s.) From earlier talk of an 'economic downturn' in the early 1970s, to the 'reces-

Table 1.2 Unemployment levels as a percentage

	1933	*1959–67*	*1973*	*1977*	*1979*	*1981*
Belgium	10.6	2.4	2.9	7.8	8.7	12.9
Denmark	14.5	1.4	0.7	5.8	5.3	9.5
France	na	0.7	1.8	4.8	6.0	8.9
Germany	14.8	1.2	1.0	4.0	3.4	6.7
Ireland	na	4.6	5.6	9.2	7.5	11.5
Italy	5.9	6.2	4.9	6.4	7.5	9.6
Japan	na	1.4	1.2	2.0	2.0	2.2
Netherlands	9.7	0.9	2.3	4.1	4.1	10.2
UK	13.9	1.8	2.5	5.7	5.8	11.3
USA	20.5	5.3	4.9	7.0	5.8	8.9

Source: Freeman, Clark and Soete (1982)

sion' of the mid 1970s and the 'recovery' of the 1976–79 period the realization has dawned that the world economy is in a state of crisis. Consider, for example, the following conclusion by a group of eminent academics drawn together by the Organization of Economic Cooperation and Development (OECD):

> It is apparent that the conditions of growth in OECD countries have been so profoundly modified as to make it unlikely, at least in the short and medium term, that rates of growth will be other than moderate. Moreover, slower growth is accompanied by high unemployment and persistent inflation. After thirty years of rapid, indeed unprecedented development, where sustained growth proceeded in step with full employment, there is now uncertainty not only regarding the rate of growth which can be achieved, but also the capacity of conventional policy instruments to reduce inflation and unemployment at the same time (OECD, 1980a, p13).

Even Samuelson, one of the foremost neo-classical economists of recent years, is drawn to a pessimistic perspective on the future of the global economy:

> No one can predict the future with confidence. Still, it is my considered guess that the final quarter of the twentieth century will fall far short of the third quarter in its achieved rate of economic progress (Samuelson, 1981a).

But what is the link between this recognition of global economic crisis and the decision of firms such as GE, Westinghouse and IBM to utilize and produce automation equipment? After all, we are frequently led to believe that it is the very diffusion of automation equipment which has, by displacing labour and hence reducing aggregate demand, in fact, caused the economic problems of unemployment and stagnation which now beset the advanced economies. In order to explain the view that automation equipment is being introduced as a response to crisis, rather than to be causing it, it is essential to go back in time to gain an historical perspective on what is now happening and what is likely to occur in the future.

A number of economic historians have recently begun to review the nature of growth over the last two centuries. Building upon a history of economic thought encompassing the writings amongst others of Marx, Schumpeter and a Russian writing in the 1930s called Kondratieff, attention has been placed on so-called long waves of economic activity, often called 'Kondratieff waves' after his pioneering work. Basically, the argument is that there are long wave cycles of economic activity and that the contemporary situation must be seen as part of the downswing of the most recent cycle. While in the past these cycles appear to have had a duration of around 50 years, it is readily acknowledged that there is no justification for any fixed period of this sort. How do we explain the existence of these longwaves of economic activity? There are a wide variety of alternative views on this and a bibliography of explanations can be found in Barr (1979) and Coombs (1982). Kondratieff, for example, believed that they were caused by the wearing out over a long time of large infrastructural investments such as railroads; Rostow (1978) puts particular emphasis on the relationship between the prices of agricultural and industrial goods; Mandel (1978) and Forrester (1976) focus their attention on the capital goods sector; there are even monetarist theories to explain the existence of these longwaves.

However the most persuasive of the various theories, we believe, are those which explore the link between long waves and technologies. These go back not just to Kondratieff, but also to Schumpeter and Marx. Recently a particularly credible theory has been produced by Freeman (1979) and Clark, Freeman and Soete (1980). Their argument runs as follows. Economic growth in the West has been fuelled by a supply-side motor, with entrepreneurs pursuing the goal of monopoly profit and achieving these profits through the introduction of new products and production technologies. Their achievements were reflected in a series of investments by competitors who were attracted by these superprofits. The result was that following the introduction of a new technology, competition increased until the monopoly profits were gradually whittled away.

But, as they point out, even if this should indeed be the motor of economic growth, there is no reason why this in itself should lead to cycles of activity. The cycles, they argue, are in part created by inflexibilities, lags and imperfections in the behaviour of both capitalists and labour, and by the fact that large investments cannot be subdivided into smaller components. Yet despite the existence of these imperfections and indivisibilities, this does not necessarily imply the existence of cycles, since a sufficient number of randomly distributed mini-cycles should even out any long-term fluctuations. Thus, they argue:

> Big wave effects could arise if some of these innovations were very large and with a long time span in their own right (e.g. railways) and/or if some of them were interdependent and interconnected for technological and social reasons (Clark et al, 1980, p25).

Freeman, basing himself on the earlier pioneering work of Kondratieff and Schumpeter, argues that over the course of the last 200 years, there have indeed been a series of major, 'heartland technologies' which have fuelled these long-run cycles ('big wave effects'). The first of these, beginning in the late eighteenth century was based upon textiles and the diffusion of the steam-engine; the second, with its onset in the mid-nineteenth century, was fuelled by the combined expansion of railroads and the diffusion of steel; the third, with its origins at the turn of the twentieth century, was based upon the motor car, electricity and the chemical industry; and the fourth has been fuelled by electronics technology, beginning with the use of the valve in the 1930s, and proceeding with the invention of the transistor in the late 1940s, the integrated circuit in 1959 and the microprocessor in 1971.

In each of these heartland-technology based cycles, there is an 'expansionary' upswing and a 'rationalizing' downswing:

> In the major boom periods [ie the upswing] new technological systems tend to generate a great deal of employment, as the form which expansion takes is the installation of completely new capacity and since the technology is still in a relatively fluid state the new factories and plants are often fairly labour-intensive. New small firms may also play an important role among the new entrants and they tend to have a lower than average capital intensity.
>
> However, as the new technology matures [ie the downswing] several factors are inter-

acting to reduce the employment generated per unit of investment . . . . Economies of scale begin to be important and these work in combination with technical changes associated with increasing standardisation. A process of concentration tends to occur and competition forces increasing attention to the problem of cost-reducing technical change (op cit p27).

The most recent cycle began after the second world war in the context of devastated European and Japanese economies and American and British economies which were largely geared to the production of military goods. The pent-up demand of consumers meant that in the early years producers were able to sell almost anything which they could manufacture. But over the years, consumers became more discerning providing scope for the introduction of new products. Freeman suggests that a considerable portion of this post-war growth was associated with the introduction of new products during the upswing of the current cycle and based upon the heartland technology of microelectronics. Although Freeman, Clark and Soete (1982) now place less emphasis on this particular link, the microelectronics sector, nevertheless, remains central. These products not only included consumer goods (such as radios, televisions and tape-recorders) but also capital goods (radar, computers and process controls), components (integrated circuits, transistors and valves) and military equipment (radars, missile controls, etc). But, continues Freeman, this expansionary upswing, associated with the introduction of new, electronics-based products began to peter out in the mid-1970s, after which the major industrialized economies made the transition to the rationalizing downswing during which the heartland technology – that is microelectronics – began to diffuse to older, more established industries, enhancing the performance of existing products and cutting their costs of production.

As we shall see, it is in this transition from the expansionary upswing of the 1950–70 period, to the rationalizing downswing of the past decade that the diffusion of automation must be understood. But to explain why automation technology developed and diffused it is necessary to pursue the link between economic crisis and automation a little further. In so doing, we focus on three major changes which occurred in the industrialized world in the quarter century after the second world war. These were the development of 'excessive' production capacity and the consequent saturation of markets, the widening diffusion of technological capability, and the growth of conflict. In examining these three sets of developments, the logic behind the decision of firms such as GE, Westinghouse, Fujitsu Fanuc and IBM to aim at the 'factory of the future' will become clearer.

It is perhaps also worth noting at this stage, that in drawing this particular link between automation and crisis we are going against the conventional wisdom which has tended to argue that the crisis in Western economies has arisen because of the diffusion of labour-saving automation technologies. In contrast, we have suggested, and will discuss further, the view that the crisis emerged first, for largely autonomous reasons. It is the competitive conditions which have characterized this economic depression which have in fact led to the rapid diffusion of automation technologies. The employment-displacing nature of these automation technologies have merely exacerbated rather than caused the descent into crisis. We will return to this point at a variety of stages as the discussion unfolds.

MARKET SATURATION AND THE DEVELOPMENT OF EXCESS CAPACITY

As we remarked the first twenty-five years after the war were characterized by a massive pent-up demand for better housing and for more consumer goods such as household appliances and motor cars. Producers faced almost unlimited demand, subject mainly to the ability to produce the products in sufficient numbers. The very production of these goods (through what economists call the multiplier effect) provided the incomes for consumers to buy them and a virtuous circle of production and growth was established.

The goods with the largest multiplier effects (eg for components, which stimulated further investments by other suppliers) were consumer durables, that is those goods which could be reused. These included 'whitegoods' (ie household appliances), televisions and motor cars. But the very 'reusability' of these goods implied that there would ultimately be a limit to their demand. Over the years this increasingly occurred with a growing tendency towards market saturation. For example, in Table 1.3 we can see the degree of penetration into UK households of major consumer durables between 1964 and 1979, a pattern repeated in almost all other developed countries. By the latter period over 90 per cent of households had vacuum cleaners, cookers, refrigerators and televisions. Almost all other durables (except dishwashers) also reached near saturation levels. In Table 1.4 we can see how the household penetration of automobiles increased over the years and by 1978 there was at least one car for every three people in all the major markets (except Japan).

The saturation of demand for final consumer goods, as we have seen, depends in large part upon the absolute limit to demand for these products. It is, however, only one side of the slowdown of growth opportunities. The other side of that coin is the growth of productive capacity by manufacturers. Thus over the years there has been a tendency for the growth of productive potential to outstrip the demand for even those products which do not face absolute limits in the same way as consumer durables, such as steel, basic chemicals and ships. From Table 1.5 we can see the growth in surplus capacity in the OECD steel industries over the decade which marked the transition from upswing to downswing of the longwave. In the OECD countries as a whole capacity utilization dropped from 91 per cent in 1970 to 71.5 per cent in 1975. Similar tales can be told of the shipbuilding, basic chemicals and other mature industries.

Of course the consumer durables shown in Tables 1.3 and 1.4 are 'traditional' products and since, by definition, all durable products faced the spectre of limited demand, we should not be surprised at the slowdown in these

Table 1.3 Household penetration of consumer durables 1964–79[1]

	1964 %	1965 %	1966 %	1967 %	1968 %	1969 %	1970 %	1971 %	1972 %	1973 %	1974 %	1975 %	1976 %	1977 %	1978 %	1979 %
White goods and major consumer durables																
Washing machine (i)	54	56	59	61	63	64	64	66	67	68	70	71	73	75	76	75
Clothes dryer (i)	17	19	21	22	24	25	25	25	26	28	30	31	31	32	34	33
Dishwasher (ii)	NA	1	1	1	1	1	1	1	2	2	2	2	2	3	3	3
Gas or electric refrigerator (i)	35	42	47	51	55	59	63	67	71	76	80	84	87	89	91	92
Vacuum cleaner (ii)	77	78	80	81	82	83	84	84	85	89	89	90	90	91	92	94
Gas cooker (i)	62	63	62	62	63	61	61	59	58	58	57	57	57	56	55	56
Electric cooker (i)	33	34	35	36	37	37	38	40	41	42	43	43	43	43	43	41
Separate freezer (i)	NA	NA	NA	NA	NA	NA	NA	3	5	8	10	13	16	20	24	25
Fridge freezer (i)	NA	NA	NA	NA	NA	NA	NA	NA	NA	NA	NA	5	8	10	13	15
Heating appliances																
Central heating system (iii)	13	16	19	21	24	28	32	34	38	40	46	49	51	54	54	56
Electric space heating (i)	74	77	78	81	81	83	83	84	86	84	83	80	75	73	70	65
Gas space heating (i)	15	19	22	25	29	31	32	34	36	37	39	41	42	44	44	45
Oil/paraffin heater (i)	31	35	36	35	33	31	29	28	28	28	26	23	22	21	21	20
Water heating (i)	85	85	85	89	93	93	96	96	97	97	98	98	98	98	97	97
Television (iii)																
Colour television	NA	NA	NA	NA	1	1	3	6	17	30	39	45	51	58	67	69
Colour and/or monochrome television	87	87	88	90	92	92	93	94	95	95	97	96	96	97	97	96
Kitchen/personal appliances (ii)																
Electric iron	93	94	94	95	97	98	97	97	96	98	98	97	97	96	96	97
Electric kettle	42	45	47	49	53	54	56	59	60	66	69	71	72	72	75	76
Electric toaster	13	15	17	17	19	20	20	20	21	24	26	28	30	32	33	34
Electric coffee maker	NA	NA	NA	NA	NA	NA	8	9	9	11	12	13	13	15	15	15
Food and drink mixer	6	8	11	13	18	22	24	26	29	34	38	40	42	44	46	47
Dry shaver	37	37	39	39	39	39	39	38	38	39	39	39	40	40	41	36
Electric hair dryer	28	32	35	37	39	41	43	44	43	48	50	54	58	59	62	63
Electric blanket	38	40	45	46	49	50	51	51	52	52	51	50	50	47	49	46

(i) 31st March (ii) 30th June (iii) 31st December NA – Not available

[1]Based upon sample surveys of over 15,000 households in each year.

Source: *Home Audit Facts and Figures*, Audits of Great Britain Ltd., Ruislip 1980.

Table 1.4 Car density in selected OECD countries, 1960–1978

	Cars per 1000 population		
	1960	*1970*	*1978*
Australia	187	314	385
Belgium	82	213	303
Canada	230	370	435
France	119	254	333
Italy	40	190	303
Japan	5	85	185
Netherlands	45	200	286
Sweden	160	285	345
UK	106	210	256
USA	344	434	526

Source: Bloomfield (1978)

Table 1.5 Production, capacity and utilization rates in the steel industry within OECD

	1965	*1970*	*1973*	*1974*	*1975*
USA					
Production	119.3	119.3	136.8	132.2	105.8
Capacity	135.5	138.7	142.5	140.6	138.9
Utilization rate	88.0	86.0	96.0	94.0	76.2
EUROPE					
EEC					
Production	113.9	137.5	150.1	155.6	125.6
Capacity	134.0	156.5	173.8	178.9	190.1
Utilization rate	85.0	87.9	86.4	87.0	66.1
Other countries (1)					
Production	13.7	21.9	26.4	28.2	26.5
Capacity	15.8	25.2	29.1	30.8	32.0
Utilization rate	86.7	86.9	90.7	91.6	82.8
JAPAN					
Production	41.2	93.3	119.3	117.1	102.3
Capacity	47.4	103.0	129.5	136.3	148.0
Utilization rate	86.9	90.6	92.1	85.9	69.1
Total OECD					
Production	302.6	390.1	453.6	454.5	381.3
Capacity			498.7	510.1	533.4
Utilization rate			91.0	89.1	71.5

(1) Excluding Yugoslavia.

Source: OECD, The Situation in the Iron and Steel Industry, Paris, 1977 (ref. C (77)104)

sectors. It is to be expected, however, that in the light of this saturation new consumer durables would be introduced which would provide substantial avenues for growth. However, here there have been two problems. First, the new products which have been introduced, (such as videos, home computers and digital hi-fi systems) have had low linkage effects. They do not require as many bought-in components as the older traditional products and they certainly do not appear to require as much labour in manufacture as, for example, did the motor car or the earlier designs of televisions. And second, as noted in the earlier discussion of long wave cycles, the phase of the new product development, especially for the consumer, was largely exhausted in the expansionary upswing. Now, as we shall see in the following chapters, the primary area for the diffusion of 'new' electronics-based products lies in the office and in the factory – in both cases these are associated with rationalization and efficiency, rather than expansion.

THE DIFFUSION OF TECHNOLOGY

In the 1950s and 1960s it was customary to talk of the technological gap between the USA and the rest of the industrialized world. (eg Servan Schrieber, 1968; OECD Gaps in Technology Series 1968, 1969, 1970) There was good reason for this as American industry tended to use much more efficient technologies producing better and more attractive products. Spurred on by management strategies with wider horizons US firms were able to take advantage of their technological leadership by investing in Europe, and selling technology to Japan which was reluctant to allow foreign investors.

As well as importing technology from the US (and to an extent from the UK), the 'backward' industrialized economies also began to devote considerably more resources to their own research and development (R & D). Moreover, as we can see from Table 1.6, since much of US (and UK) R & D was centred on the military sector, the spin-off to industrial sectors in other industrialized economies was greater, thereby closing the technological gap with the more advanced American and British economies. For example, in an interesting set of readings (Pavitt (ed), 1981) various observers argue that the preponderance of UK, R & D in the defence, aerospace and nuclear industries directly contributed to its falling share of innovations and world markets in manufactured goods.

Consequently right through the post-war period productivity growth in the 'backward' industrialized economies was considerably faster than that in the USA and the UK (Table 1.7). The technological gap was beginning to close and this had two main consequences. First, Japanese and European firms began to grow more rapidly than US firms (Rowthorn and Hymer, 1971). For example, whereas the US held first and fourth spot in the world's largest chemical companies in 1970, by 1980, the first three places were held by German companies with the American counterparts slipping to fourth and sixth place

Table 1.6 Trends in expenditure on R & D as a percentage of GDP in selected countries
Total, defence and other

	*1963**	*1967*	*1971*	1975
Canada				
Total	1.00	1.20	1.20	1.00
Defence	0.09	0.09	0.06	0.04
Other	0.91	1.11	1.14	0.96
France				
Total	1.60	2.20	1.90	1.80
Defence	0.43	0.55	0.33	0.35
Other	1.17	1.65	1.57	1.45
Germany				
Total	1.40	1.70	2.10	2.10
Defence	0.14	0.21	0.16	0.14
Other	1.26	1.49	1.94	1.96
Italy				
Total	0.60	0.70	0.90	0.90
Defence	0.01	0.02	0.02	0.02
Other	0.59	0.68	0.88	0.88
Japan				
Total	1.30	1.30	1.60	1.70
Defence	0.01	0.02	–	0.01
Other	1.29	1.28	–	1.69
Netherlands				
Total	1.90	2.20	2.00	1.90
Defence	–	–	0.04[a]	0.03
Other	–	–	(1.96)	1.87
Sweden				
Total	1.30	1.30	1.50	1.80
Defence	0.40	0.43	0.23[b]	–
Other	0.90	0.87	(1.27)	–
United Kingdom				
Total	2.30	2.30	2.10	2.10
Defence	0.79	0.61	0.53	0.62
Other	1.51	1.69	(1.57)	1.48
United States				
Total	2.90	2.90	2.60	2.30
Defence	1.37	1.10	0.80	0.64
Other	1.53	1.80	1.80	1.65

*Germany, Netherlands, Sweden, United Kingdom, 1964.
[a]1972.
[b]1970.
Source: OECD, 1980a, p 30

(Betts, 1981). And second, instead of the flow of foreign investment being overwhelmingly from the USA to the rest of the world, it began to become increasingly complex. European firms came to interpenetrate each others' markets, becoming continental firms rather than national firms, and both European and Japanese firms began to make inroads into the American market undertaking significant investment in the USA. There were two spurs to this

Table 1.7 Annual productivity growth in manufacturing in selected countries (percentages)

	1955–60	*1960–4*	*1964–9*	*1969–3*	*1973–7*
France	4.9	5.1	5.2	5.1	2.4
Germany	5.0	4.8	4.9	4.2	3.2
Italy	4.6	6.3	6.2	4.8	1.0
Japan	7.3	10.3	8.8	8.1	2.7
Netherlands	3.6	3.4	4.9	4.5	2.7
UK	1.8	2.3	2.6	2.9	0.4
USA	1.0	3.1	1.9	1.5	0.3

Source: OECD, 1980a, Table 13.

growing interpenetration of investment. The first largely affected the entry of both European and Japanese firms into the USA, arising in the late 1970s when the US dollar declined in value and foreign firms took the opportunity of taking over established US firms cheaply. Second, at about the same time, and largely against their wishes, Japanese firms were forced to invest in local production in both Europe and North America to circumvent the barrage of import controls which arose directly out of their success in selling Japanese-made goods in these markets.

While the major factor allowing this interpenetration of markets was the increasing spread of technology through industrialized countries as the upswing of the longwave cycle began to mature, a second factor which made this possible (and in fact inevitable) was the growing tendency towards overcapacity in each of these countries. The fact that local markets were growing more slowly forced firms to become transnational in their approach, ensuring growth in countries where demand was expanding more rapidly, where they could make inroads into the markets of competitors and where they could take over established companies.

THE GROWTH OF CONFLICT

In completing our review of developments in the most recent long wave cycle it is essential to bear in mind the increasing trend towards conflict within the system. There are two primary areas of conflict which interest us in explaining the drive towards automation. These are the conflicts between capitals, and the conflict between capital and labour.

The conflict between capitals

One of the most interesting facets of the post-war global economy is the unevenness of development. This is reflected by the relatively slow growth rate (of output and productivity) of the USA and UK; the exceptionally strong performance of Japan and the relatively strong performance of Germany; and the imbalance between the size of individual European economies and the USA. At the same time there has been a tendency for firms to become transnational (that is a trend towards concentration of ownership and production at a world level) and for their growth to exceed those of firms which remained predominantly national in their orientation. This unevenness, associated with a growing interpenetration of markets – initially through trade and subsequently through investment in direct production – has provided the scenario for a growing measure of competition and conflict. This has taken a variety of forms, differing over time and between sectors. First, in the production of 'traditional' products, there has been a tendency towards conflict between national and international firms within each country. For example, in the UK the latter half of the 1970s saw the favouring of the interests of the international fraction of capitalists, those firms with a relatively large proportion of production and sales taking place outside the UK: entry into the Common Market made it easier for them to sell and invest in Europe, and increased the competitive pressures on local British firms. Then the advent of the monetarist government in 1979 with the consequent rise in the value of sterling (by over a third within two years) substantially tightened the screw: the international fraction of capitalists took advantage of the strong pound to export their capital abroad, and local manufacturing industry was decimated (with a loss of 15 per cent of all jobs in manufacturing industry in 15 months!). These developments have been especially marked in the engineering goods sector, where batch production has not hitherto allowed for the large scale-economies which at an earlier stage forced the internationalization of production in other mass production sectors.

Second, there has been a growth in competition between different national capitals. The growth of state sponsorship of Japanese and French firms has been a particularly marked phenomenon, but this alliance between state and corporation, has occurred in almost all countries. It involves direct state subsidies, preferential state purchasing and intergovernmental bargaining on behalf of national firms. This alliance between the state and national capital has been particularly prominent in the new, emergent high technology sectors such as telecommunications and electronics, in the defence-related industries and in declining traditional industries such as steel and shipbuilding.

Third, there has been a growth in competition between international capitals, most notably between European, Japanese and American capital. Thus, for example, in the motor industry, in the machine tool industry and in the textile industry, it has been the EEC, representing an alliance between various European capitals, which has been conducting negotiations with American and Japanese firms.

It is not possible to determine which particular one of these sets of conflicts has predominated, since this has varied over time and between different sectors. But, as a general observation, we can note that the past developments of unevenness and concentration are tendencies endemic to the system and that we can expect them to continue in the future. Moreover, because of the prevalence of excess capacity, declining growth rates and the diffusion of technology, we can also anticipate that unless governments intervene to impose 'orderly marketing agreements', the intensity of this competition will increase even further. It is appropriate, therefore, to characterize the current crisis as a state of 'supercompetition' with complex and changing alliances between different fractions of capitalists and between these fractions, the state and supra-state organizations.

The conflict between capital and labour

Referring back to Table 1.2 it is evident that the 1950s and 1960s were periods of reasonably full employment (albeit with declining success) in most industrialized economies. In particular, despite the existence of a reservoir of unskilled unemployed during this period (particularly in the peripheral European and Carribean economies) which provided a ready source of immigrant labour, semi-skilled and skilled workers were in short supply. The consequence was a growing trend towards unionization of the labour force and the heightened emergence of industrial disputes. (Crouch and Pizzone (eds), 1978) Furthermore real wages continued to rise right through the period.

To summarize, therefore, as the longwave cycle progressed through the expansionary upswing and matured into the recessionary downswing, capital

Fig. 1.1 Profit rates in manufacturing

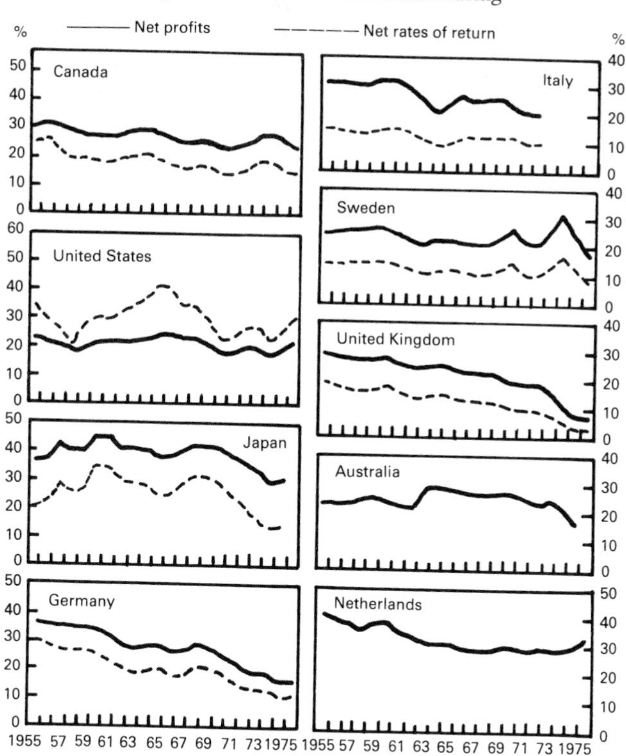

Source: Hill, 1979

was squeezed at both ends: in the final market by a growing phase of super-competition and in production by the growing power of labour. Moreover, the increasing capital intensity of production heightened the financial requirements for entering production in most sectors. The consequence, as we can see from Figure 1.1 was an inevitable falling off in the rate of profit, a phenomenon which occurred at a progressively faster pace in all of the major industrialized economies.

THE DRIVE TO AUTOMATION

Confronted with these twin problems of a growing threat to its authority and declining profit rates, capital has taken three major steps in self-defence. Ironically each of these, as we shall see, has only had the effect of worsening the crisis in the system.

The first of these, occurring first in the late 1960s and then throughout the 1970s, was a change in the labour process, that is the nature in which capital organizes its labour. The form which this took was to redefine the work involved in manufacture and assembly, and to have the labour-intensive parts of it undertaken in developing economies with a pliant labour force and low wages. In many sectors, and in particular the electronics sector, subsidiaries were opened in newly industrializing countries such as South Korea, Taiwan, Hong Kong, Singapore, and Mexico, in which part of the manufacturing processes, especially assembly, was undertaken. Interestingly it was predominantly British and American firms, who were being challenged by the growing competition of European and Japanese firms, which took greatest advantage of this decomposition of the labour process and the resultant 'offshore' assembly operations.

This change in the labour process was not confined to the switch to low-wage subcontracting but also covered the organization of production in the home countries. Faced with growing unemployment in the external economy and the threat of enforced redundancy, labour became more compliant at a general level, as well as within individual factories, with a change in work practices which allowed for a growth in productivity and a growing intensity of work. The following quote from an article in the *Financial Times* ('Some workers have already accepted wage cuts', 15 September 1981) shows the extent of this change in the balance of power.

> Wage cuts have already been accepted by some workers and the idea is likely to spread, according to Mr. Reg Parkes, vice-chairman of the Regional Council of the Confederation of British Industry.
>
> He points to the experience of the 1930s when employees took a cut in living standards, in order to preserve jobs, and he argues that the current recession is likely to prompt a similar response. Most deals were already being struck at below 5 per cent. Many companies were deferring any increase until business prospects improved.
>
> The cost of converting the once highly paid Midland workers to moderation, has been

high. In a wave of redundancies and closures, jobs have disappeared at the rate of around 1,000 a week, over the past 18 months. Unemployment has more than doubled to 14.8 per cent. Another 100,000 workers remain in employment only because of the benefits their companies claim under the temporary short-time working compensation scheme – and the CBI has warned that more redundancies are on the way.

Major Peter Forrest, chairman of the West Midlands Engineering Employers Association, talks of a new mood of realism. 'The value of a job is now appreciated. Workers know that if they get a silly wage rise jobs will go.' He maintains that most employers have reacted responsibly to their improved bargaining power in the realisation that to take advantage now would invite trouble in any economic upturn.

The shift of negotiating power to the employers, is dramatic. In the words of one regional union official, 'to say that our position has been undermined is something of an understatement. The number of disputes has dropped considerably. Workers are demoralised and know there is no point in industrial action.'

In many cases this strategy of changing the labour process proved to be successful for particular firms, but it ran into increasing problems as the decade of the 1970s grew to its close. This was because the consequence of these defensive responses by a myriad of firms was the loss of jobs in their own countries, which was part of the reason why the unemployment rates in home countries grew over the years. Faced with the loss of jobs the unions in these home countries reacted by pressurizing for import controls against these low-wage products from developing countries, and these pressure for import barriers have become increasingly intense. At this level, therefore, capital's strategy of changing the labour process became increasingly challenged, directly as a result of its success. But the strategy faced contradictions at another level as well, in that the loss of jobs in the home countries – through the multiplier effect – began to depress the growth of demand in these economies and, moreover, it had budgetary consequences for the state which has had to finance unemployment through increased welfare costs. This has meant higher interest rates which both dampened demand further and meant competition between the state and capital for scarce savings.

The second response for capital has been at a macro level and has necessitated action on a broader political front. Confronted with the growing militancy of labour, capital has pressed for tougher action, at the level of the state, against labour. Added to this (as we have seen) has been the erosion of worker militancy due to the presence of a growing army of unemployed. Thus, towards the end of 1980 we have begun to see a reduction in real wages in some countries, especially the USA and the UK. At first this took the form of keeping earnings in line with price rises (which because of tax, meant that real wages declined), but in the very recent period we have begun to see more significant reductions. For example, Pan American Airways, faced with growing competition on both domestic and foreign routes has enforced a wage freeze on all employees despite double-digit inflation. Other companies have followed this lead and in yet other cases (eg the Hoover Corporation in the UK) there have actually been cuts in nominal wages.

This strategy of confronting labour, enforced by wider monetarist policies which increased the size of the army of unemployed and cut the social wage (eg in welfare benefits), holds dangers for the capitalist system since it ultimately restricts demand. This as we have already seen, was one of the major problems with which the system was faced over the 1970s and can only act to increase the severity of the recession in the 1980s.

The final and most recent response of capital to the growing intensity of competition has been to introduce automation technologies which, as we shall see in subsequent chapters, offer the benefits of lower unit costs and better quality products produced with a shorter lead time. Consider, for example, the case of the automobile industry where Japanese producers have penetrated US and European markets with a plethora of low-cost, reliable models. In each country the automobile industry has induced their governments to introduce formal and informal barriers to the numbers of Japanese car exports. Attempts have also been made to obtain the 'cooperation' of the trade unions to settle for lower wage rises and agree to changes in work practice which increase the intensity of work. But perhaps most significantly, all of the major automobile producing firms have committed themselves to the very rapid introduction of automation technology. Not only has this meant significant investments but in many cases (eg Volkswagen, Fiat and Renault) the automobile firms have been forced to manufacture the automation equipment themselves, since no other domestic suppliers have been able to respond with appropriate urgency.

Here we can clearly see the link between economic crisis and automation. The 1960s and 1970s saw the faltering of global economic expansion and the emergence of depression. Associated with this was a markedly more competitive economic environment. In this environment the path to survival for individual firms is tortuous. It requires the help of governments in fending off competition and financing regeneration. It also requires a renewed assault against the militancy of labour which had grown powerful during the years of economic expansion and near-full levels of employment. But, most pertinantly, it has forced individual firms to introduce automation technologies. Without doing so, it will be almost impossible to supply competitively priced products, of adequate quality and within acceptable time frames. Thus although it is indeed possible that the emerging depression is in part explained by the diffusion of automated technologies, the more powerful causal explanation turns this relation on its head. It is this phenomenon which we address in the coming chapters.

CHAPTER TWO

Forms of Automation

In his survey on the literature of automation, Bell (1972) concluded that:

It quickly became apparent that the existing literature on 'automation' was woefully inadequate. There is a very large body of literature, but there is no agreed set of concepts. The mass of empirical case-study material, not surprisingly, lacks structure and consistency of results (p58).

Today, the situation is little changed. The problem is that we are presented with two definitional extremes. At the one end is a generalized view of the concept of automation which tends to be that used in everyday conversation. For example, the Oxford English Dictionary defines it as:

Automatic control of the manufacture of a product through succesive stages, (loosely) use of machinery to save mental and manual labour.

This is a perspective shared by a range of other observers. Einzig (1957), in a well-known earlier text, defines it as:

a technological method that tends to reduce current production costs in terms of man hours per unit of output . . . Its loose use practically as a synonym for advanced mechanisation may shock the technologist, but serves the purpose of the economists (p2)

At the other extreme are observers who concentrate on the specific ability of automation technology to control particular activities, an ability usually associated with human beings. Hence

. . . 'automation' is a technology quite distinct from 'mechanisation' and it is concerned with replacing or aiding human *mental* effort as distinct from aiding man's physical effort (Thomas, 1969, p6).

Underlying this is the concept of feedback control (as represented in Figure 2.1) which involves three distinct sub-processes: the ability to inspect and measure (or 'sense' as it is commonly known), the evaluation of this measurement in relation to a theory (or algorithm) of the process, followed by some form of reaction if a response is required. It is this particular form of automation which was christened with the name cybernetics (literally 'the art of the

Fig. 2.1 The feedback (or cybernetic) view of automation

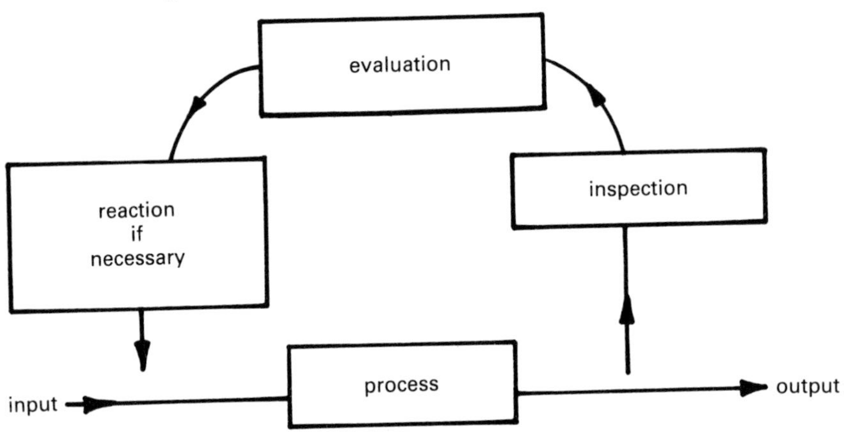

steersman', from the Greek *kybernetes* steersman) by an influential American mathematician Norbert Wiener, in 1947.

It is difficult to reconcile the cybernetic view of automation with that which encompasses advanced forms of mechanization since they represent two different types of activity. On the one hand the definition which requires a feedback mechanism is intuitively appealing since it imples some form of cognitive system – an automatic response. Yet despite this appeal it would be misleading to limit the definition of automation in this way since it would exclude, for example, a highly mechanized (albeit inflexible) mass production line in which components are fed in at one end and emerge as assembled items at the other, without either human intervention, or machine controlled feedback being involved.

Hence there is some state of confusion in the literature and it our task in this chapter to try and provide some shape to the concept of automation. In doing so, it is necessary to begin with the widest possible perspective, embracing both feedback and mechanization. This generalized view offers little possibility for analytic distinctions and as Bright (1957) – who is probably the most celebrated observer of automation – concluded, the only acceptable categorization which this wide perspective allows is one of degree, that is

> something more automatic than previously existed in that plant, industry or location (p6).

This categorization of automation allowed Bright to develop seventeen levels of mechanization (see Figure 2.2). It is interesting to note that all levels beyond the second explicitly involve some degree of control.

A similar approach is taken by Amber and Amber (1964) who with Bright, made an important early contribution in charting the development and progress of automation technology. They, too, see the phenomenon as best represented in terms of degree, developing ten orders of automaticity (see Table 2.1). The characteristic of all these orders is that

Fig. 2.2 Seventeen levels of mechanization and their relationship to power and control sources

Initiating control source	Type of machine response		Power Source	Level Number	Level of mechanization
From a variable in the environment	Responds with action	Modifies own action over a wide range of variation	Mechanical	17	Anticipates action required and adjusts to provide it.
				16	Corrects performance while operating.
				15	Corrects performance after operating.
		Selects from a limited range of possible pre-fixed actions		14	Identifies and selects appropriate set of actions.
				13	Segregates or rejects according to measurement.
				12	Changes speed, position, direction according to measurement signal.
	Responds with signal			11	Records performance.
				10	Signals preselected values of measurement. (Includes error detection)
				9	Measures characteristic of work.
From a control mechanism that directs a predetermined pattern of action	Fixed within the machine			8	Actuated by introduction of work piece or material.
				7	Power tool system, remote controlled
				6	Power tool, program control (sequence of fixed functions).
				5	Power tool, fixed cycle (single function).
From man	Variable			4	Power tool, hand control.
				3	Powered hand tool.
			Manual	2	Hand tool.
				1	Hand.

Source: Bright (1958)

Table 2.1 The ten orders of automation

Order of automation	*Human attribute mechanized*	*Some examples*
1. Hand tools and manual machines	None	Shovel, knife, handloom.
2. Powered machines and tools	Energy	Electric drill, spray gun.
3. Single-cycle automatic machines	Dexterity	Grinder, lathe, radial drill.
4. Repeating cycle machine	Diligence	Engine production line, copying lathe, non self-correcting NC and record-playback machines.
5. Self measuring and adjusting machine, incorporating feedback	Judgement	Process controllers, pattern-tracing flamecutters, self-correcting NC machinery.
6. Automatic cognition, computer controlled	Evaluation via programmed algorithm	Selective assembling robots, computer-aided design.
7. Limited self-programming	Learning	
8. Relates cause from effects	Reasoning	Sales prediction, weather forecasting.
9. Originality	Creativeness	
10. Commands others	Dominance	

Source: Adapted from Amber and Amber (1964) pp2–3.

whenever a machine assumes a human attribute, it is considered to have taken an 'order' of automaticity (p3).

Where Amber and Amber differ most from Bright is in regard to the ultimate level of automation envisaged. Amber and Amber's sixth order proximates to Bright's seventeenth level, whereas their highest order ('commands others') is hopefully fantastic and in the realm of imagination rather than reality.

However useful these early contributions were in providing some form of order to a chaotic field, they had one crucial weakness in that they failed to distinguish clearly enough between different forms of automation. They concentrated only on its degree. It was Bell (1972) who first clarified the key components of automation technology. He questioned the focus of Bright and others who had hitherto suggested that automation involved a hierarchical progression of control. To the contrary, argued Bell, control was only one of three distinct components of automation, the others being the transformation of inputs and their transfer between workpoints. In each of these areas there exist degrees of automation, but a high level in one area need not necessarily be associated with a high level in the other two.

Now this was an important insight since for the first time it provided for a difference in the types of automation as well as a difference in its degree. Bell (and more recently Coombs, 1982) went on to argue that advances in these types of automation technologies occurred at different periods of history. The automation of transformation began first, in the eighteenth century (with the introduction of water power), developing further in the nineteenth century (with steam power) and the twentieth century (with the introduction of electricity and internal combustion engine). In each case complementary advances in materials technology (for example, high-speed steel) further enhanced the degree to which transformation machinery became more productive. Then, towards the end of the nineteenth century, these advances in transformation technologies came up against the bottleneck of transfer, and the need to speed up the whole operation rather than merely that of transforming particular inputs. This involved not merely automation of transfer itself (for example, the use of conveyor belts) but, perhaps more significantly, a reorganization in the way in which production occurred. 'Scientific management' (of which we will hear more in Chapter 8) and the assembly line were perhaps the key organizational outcomes in the 1875–1925 period. Finally, it is in the most recent period that the automation of control has become most marked, both because of the need to make increasingly productive transfer lines more flexible, and because of the extent to which emerging technologies (especially electronics) facilitated this flexibility. It is perhaps not surprising therefore, that it was in this period that Wiener and others, through their emphasis on feedback control, sought to institute this as the 'true' characteristic of automation in general.

This, then is the current state of the art with regard to the characterization of automation. It comprises three dimensions, First, automation should be considered in its widest sense; second, there are degrees of automation; and third, automation consists of three components, namely, transformation, transfer and control. But is this adequate? Does it give us a sufficient hold on the concept to explain the nature and significance of developments which are now unfolding and which will give us the 'factory of the future' as promised by GE, Fujitsu Fanuc and their competitors? We believe not, and the reason is that this literature only covers one – albeit a key – sphere of production, namely the physical transfer of inputs into outputs. But what of the other important technological developments which are now unfolding, such as those in the office, and those in the conception and design of new and improved products? These, too, are important elements in the organization of production and lend themselves to analysis in a similar way. In order to understand the significance of this critique of the state-of-the-art studies on automation, it is necessary to offer first a brief description of the organization of production in the modern enterprise.

In the modern industrial firm, as we can see from Figure 2.3, there are essentially three *spheres* of production. The first of these is design where the nature of the firm's output (eg cars, buildings, sweets) is defined and new production processes are explored. The key actors in this sphere are skilled

Fig. 2.3 Pre-electronic organization of factory production

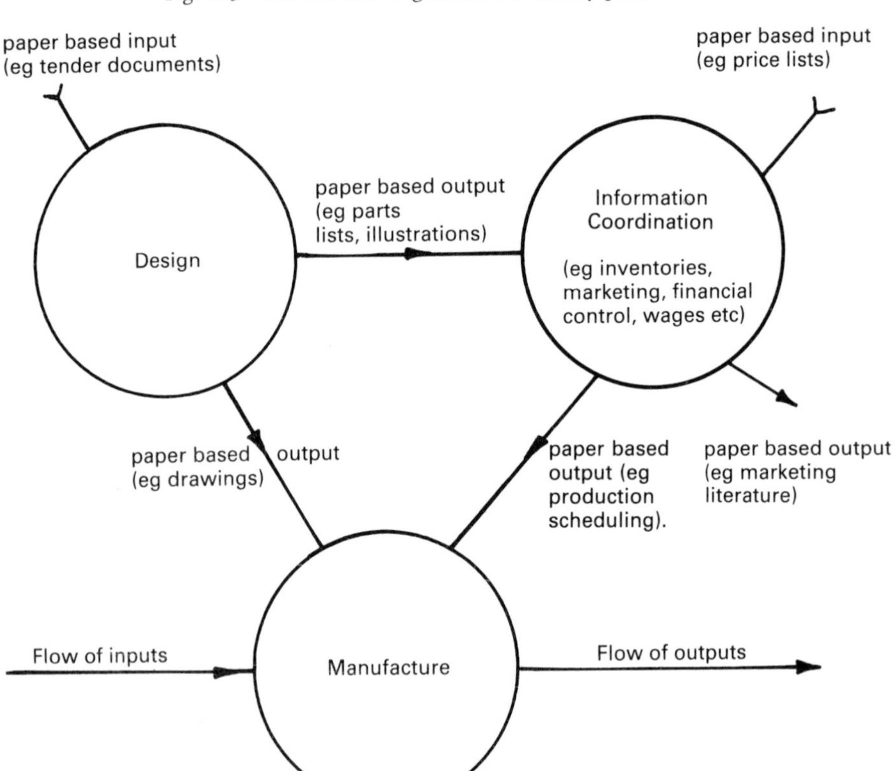

engineers, scientists and technicians, but to work effectively they require the back-up help of a staff of 'information processing' assistants such as secretaries and librarians. The actual manufacture of these designs occurs in a second sphere of production in which the raw materials and intermediate inputs are stored, processed into final products and ultimately delivered to the consumer (which is often another affiliate of the same firm). These two spheres of production, which are the kernel of an enterprise's activities, could not operate effectively without some form of coordination and this comprises the third sphere of production.

Naturally the extent to which these spheres of production exist in any particular enterprise depends upon the nature of the activity involved. Firms producing simple products with relatively 'low technology' will obviously have a poorly developed design function, whereas small, high-technology electronics firms may have a very well-developed design function which requires little formal coordination. Nevertheless in almost all modern enterprises, whatever their sector or size, these three spheres of production will tend to be separated into different units (often in different towns, cities and even countries): the R & D block, the factory and the administration. Clearly this separation of function has not always been the predominant form of organization; a point to which we will return later in this chapter.

Now within each of these three spheres of production there are a variety of separate *activities* (which are discussed in greater detail in Chapter 3, 4 and 5). For example, within the design sphere, design itself is usually an activity distinct from drawing, copying and tracing; within the manufacturing sphere there are important differences between handling, forming, assembling, control, storage and distribution; and within the coordination sphere, information has to be gathered, processed and stored. Whilst some activities are common to all enterprises – for example, handling in the manufacturing sphere – there will inevitably be a variation in the number and type of other activities. This variation is particularly marked in the manufacturing sphere, where it will be affected by factors such as the nature of the process (flow or batch) and scale (small or large batch).

Armed with the recognition that there are these three spheres of production, each with its particular sets of activities, it is possible to categorize three different types of automation. We will briefly describe these three types of automation and will treat them in greater detail when we focus on each of the spheres of production in coming chapters.

Intra-activity automation refers to automation which occurs within a particular activity. Clearly, in line with our earlier definition of automation, this intra-activity automation may take a variety of forms ranging from the simple substitution of machine power for human power (as in the use of computer-aided draughting systems) to the more complex incorporation of machine 'intelligence' and control (as in computer-aided design systems). The determining characteristics of this type of automation, however, is that it is limited to a particular activity and that it is consequently isolated from other activities within or beyond, the particular sphere of production. In many of the earlier studies of automation, this is referred to as 'island' automation.

Intra-sphere automation refers to automation technologies *which have links with other activities within the same sphere*. Indeed, the origins of the term 'automation' in the Ford assembly plant of the 1920s illustrate this type of automation well: the new transfer line mechanized the flow of materials between different activities such as lathes, drilling and boring machines. In its more complex form – as in the newly developing flexible manufacturing systems – intra-sphere automation involves the monitoring of the progress of production with an ability to adjust components of individual activities, if this becomes necessary.

Inter-sphere automation is the third and most complete form of automation and *involves coordination between activities in different spheres of production*. In view of the number of activities within each of the different spheres, there are a wide variety of potential inter-sphere combinations. These may be of a relatively limited and simple nature, for example using design parameters to automatically set machine settings; or they may be wide ranging and complex such as in the linking of changes in the specification of products to paramenters generated in redesign, and thus in continual adjustments made in machine settings.

The essential difference between these three different types of automation is shown in Figure 2.4. In (a) we illustrate the introduction of automation technologies into individual activities within each of the three spheres. As we can see there is no link between these individual intra-activity automation technologies and other activities, even within the same sphere. In (b) we illustrate how an automation technology is introduced into a particular sphere with some form of interlinking (involving feedback in the case of the manufacturing sphere) between different activities. Finally in (c) we give an example of the merging of the three-sphere industrial enterprise back towards the single-sphere type of organization which as we shall see characterized pre-industrial revolution enterprises. In this case automation technology links up different activities between different spheres of production.

In the same way that the three components of automation – namely transformation, transfer and control – evolved over the last three centuries, so too has the shape of the enterprise, and consequently, the three types of automation. Consider for example, the following description of 'traditional' economies:

> . . . there is no society without its specialists such as weavers, potters or woodworkers. These differ from the specialists of our industrial society in that they do not get their subsistence by exchanging their products for food and shelter. Every craftsman is a food-getter as well. He cannot 'earn his daily bread' because every household is expected to provide its own food. Craftsmen work in their 'spare time' – when there is no fieldwork or herding or hunting to be done, and other people would be making the rough objects that anyone can make, or just sitting around drinking beer and talking. They rarely accumulate stocks in the hope that somebody will want them, display them in front of their houses or hawk them around; mostly their products seem to be 'bespoke' (Mair, 1966, p147).

Fig. 2.4 The three different types of automation

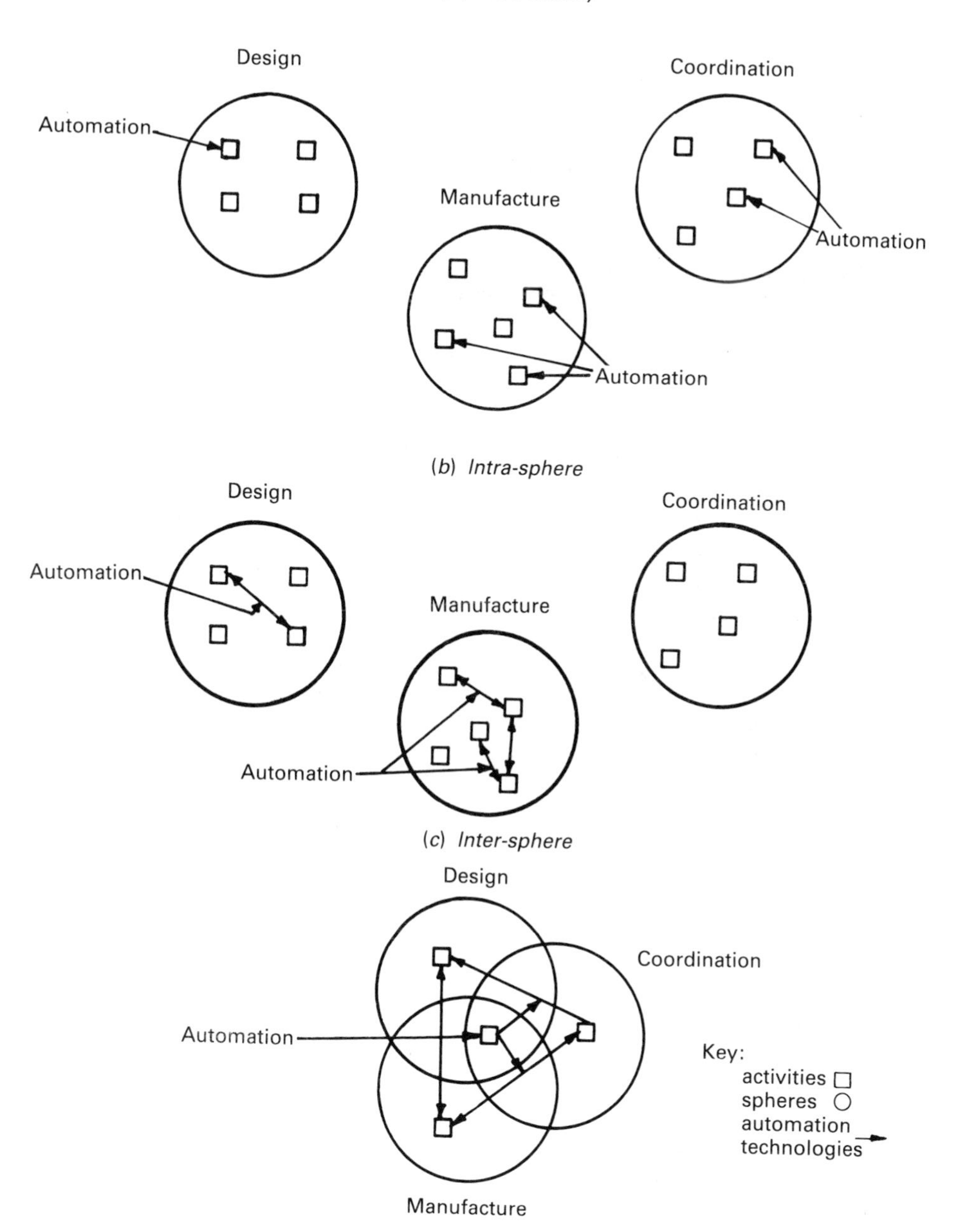

Here there appears to be little specialization, certainly not within an institution such as the enterprise. The design of products, their manufacture and the coordination of the whole process is the task of the same individual; there is a unity in the three spheres of production.

There are many cases in which this manufacture for self-consumption on a part-time basis occurred. But it seldom existed in a comprehensive form, for two major reasons. Firstly, it is seldom true that 'subsistence' units produced no surplus. Clearly there were periods of drought or natural disaster when resources were severely constrained and basic survival was at issue, but more often the basic producing unit almost always produced a margin of surplus (sometimes spare labour-time, but usually produce) which could be exchanged for agricultural implements, basic commodities or industrial services (eg roof thatching) provided by others. And secondly, the production of these industrial commodities more often than not required particular skills which imporved through practice and specialization.

The result of all this was the development of the craftsperson, who was devoted to the production of a single commodity or a limited number of similar commodities. These were produced on a full-time basis and exchanged for the agricultural output or industrial items provided by other specialized units. Sometimes this craft specialization was 'voluntarist' in that anyone cound enter the trade; in other cases, particularly in India where it became formalized into a hereditary caste system, the craft-role became very clearly defined.

At a later stage of history (for example in mediaeval Europe), craft skills became institutionalized within craft guilds, whose

> . . . essential function was to protect the artisan, not only from external competition, but also from the competition of his fellow-members. It reserved the town market exclusively for him, closing it to foreign products, and at the same time it saw that no member of the profession grew rich to the detriment of others. It was on this account that more and more minute regulations governed a technique which was strictly the same for all, fixed hours of work, settled prices and wages, forbade any kind of advertisement, determined the number of tools and of workers in the workshops, appointed overseers charged with the most meticulous and inquisitorial inspection – in a word, contrived to guarantee to each of its members both protection and at the same time as complete an equality as possible (Pirenne, 1971, p187).

These craft guilds differ from the 'traditional' societies described earlier in that there is specialization of labour and skills. However, their similarity with the traditional societies lies in the continued unity of the three spheres of production.

The next major development was the grouping together of artisans in the first capitalist enterprise, relying on the same skills which characterized the craft guilds, but putting them to work under the same roof. Here we see the first attempts to separate out the spheres of production. Design and coordination were usually the function of the capitalist entrepreneur (although the artisans often were responsible for improvements in product and process), whilst manufacture was undertaken by the employed artisans and unskilled workers.

This type of enterprise was an increasingly common phenomenon in Europe in the seventeenth, eighteenth and the first half of the nineteenth centuries. In was undermined by the development of machines and the factory system. Here instead of production being organized by grouping together artisans, machinery was introduced to supplant these artisan skills. Machines, initially arising out of improvements made during the production process itself, became increasingly complex and their design came to be a specialized task. Over the years particular firms emerged to produce machinery, but whether produced within the user-industry or by specialist producer-firms, the design task came to be separated from coordination and manufacture. As we have seen, at the same time the development of transfer-automation technology required a reorganization of production and the coordination sphere was subject to increasing specialization with the emergence of various tiers of management.

By the last quarter of the nineteenth century, therefore, the larger enterprises in Western Europe and North America had seen the evolution of the three spheres of production. The sphere of design was increasingly based upon the application of scientific principles, coordination saw the emergence of tiers of management and manufacture was characterized by the application of ever more complex machinery, involving a steady growth in the division of tasks. The last three-quarters of a century has seen the extension of this differentiated enterprise from the local to national markets, and thereafter from national to international markets. The three spheres of production continued to extend over this period until today we can observe their functioning across the globe. As the extensive literature on the multinational corporation (MNC) shows, design, coordination and manufacture in a single firm often extend over national boundaries: design and senior management in the home-country with manufacture and elements of coordination spread over a great number of countries.

What is at issue now is the transition to the automated enterprise. Whereas the last three centuries have seen the gradual evolution and specialization of the three spheres of production, what we are now beginning to witness is the re-emergence of the unitary, undifferentiated firm. The development of the automated enterprise, embodying the extension of inter-sphere automation throughout the firm, is leading once again to the unity of spheres, as illustrated in Figure 2.4. This is the significance of drawing out the three types of intra-activity, intra-sphere and inter-sphere automation. Merely focusing on its components – that is transformation, transfer and control – ignores the central importance of these emerging developments in firm structure and organization.

The aim in the following chapters is to show the unified nature of the automated 'factory of the future', and to assess the speed of its development. But before this can be done, it is important to go back to the earlier discussion of long wave cycles, heartland technologies and microelectronics. There we observed that the downswing of the current long wave cycle is characterized by the diffusion of the heartland, microelectronics technology to existing industries to enable them to rationalize production and cope with the emergence of supercompetitive pressures. One of the reasons why microelectronics is fulfilling this role is that it reduces all decision and numerical systems to binary logic, and by linking these together, facilitates the development of all three types of automation, especially inter-sphere automation. In order to comprehend the significance of the role being played by this heartland technology it is essential to discuss the technological underpinning of the automated factory in a little more detail. In doing so we concentrate on developments in control technology since these are providing the key to the emergence of full, inter-sphere automation.

Referring back to Figure 2.1 it is evident that there are four subcomponents of such feedback systems, namely the ability to inspect, or sense, processes; the ability to evaluate these processes in relation to a theory of what is occurring, that is the algorithm; the ability to activate the feedback mechanism; and, finally, the ability to communicate between these three subprocesses. Of these the least problematical has been the development of reactivation technologies and a wide variety of switching devices are now available. Sensing technologies have created greater difficulties and often constrain the advance of automated technologies. For example, one of the greatest constraints to automated bottle production remains the identification of flawed containers. The current practice is for people to examine each bottle individually in front of a neon light; aside from the manual nature of this task it is also very unpleasant and workers interchange these tasks at hourly intervals. Indeed despite the fact that there are a very wide variety of alternatives – Thomas (1969) for example lists 66 different types – the absence of suitable sensing technologies often hinders the introduction of automated systems. When suitable advances are made, the relaxation of the constraining bottleneck leads to very rapid diffusion. For example, following the initial development of sensors in the 1940s and 1950s, diffusion was largely confined to the aerospace industry, engineering laboratories and heavy process and power generating industries. But in the late 1970s two crucial breakthroughs were made, namely the mass production of low-cost sensors for automative control and the linking-up of sensors to microprocessors. The resultant growth of the industry was phenomenal; the market in the US alone in 1981 for 5 million sensors exceeded the total sum of world demand for all sensors in all the preceeding years (Lund et al, 1980). Despite these significant recent advances, the absence of suitable low-cost sensing technologies remains a constraint to the development of automation in many sectors.

But the key to development in the advance of automation has been the recent emergence of informatics technologies, that is the link between electronic processing and communication technologies. In order to grasp the significance which electronics has to play in this area of control and communication it is first necessary to understand the process of binary logic. Binary (often called digital) systems operate on the basis of an either/or logic in which counting and logical systems are decomposed into a variety of stages, each of which can be answered with binary logic. For example, in Figure 2.5 we illustrate how with the use of a series of switches (often called 'gates' or 'bits') it is possible to count. Each of the switches can be 'on', in which case its value is recorded, or 'off' in which case it is not. By so doing any number can be calculated.

Fig. 2.5 Binary logic in counting system

Value of switch	1	2	4	8	16	32	64	128	256	512	Number counted
'on'	x		x						x		261
'off'		x		x	x	x	x	x		x	
'on'	x										1
'off'		x	x	x	x	x	x	x	x	x	
'on'	x							x		x	641
'off'		x	x	x	x	x	x		x		

Similarly, with the use of this identical binary classification, the operating logic of control systems can be defined; this includes trade union response to the introduction of office automation equipment (Figure 2.6).

In the early years this binary logic was performed by thermionic valves (as in the old 'valve-radios'). These formed the basis of the first digital computers which were developed in the 1940s to calculate the trajectories of shells fired from artillery guns. The problem with these valve-driven systems was that they were cumbersome, unreliable and used great quantities or energy. But the development of the transistor in 1947 established a technique for providing binary logic gates which had no moving parts and which instead were based upon the interrupted flow of electricity. Consequently the transistor was inherently cheaper and more reliable than the thermionic valve. Subsequently two critical developments, namely the integrated circuti which was developed in 1959 and which contained more than one logic gate on each component, and the programmable microprocessor developed in 1971, which provided the capability to program separate integrated circuits, speeded up the diffusion of solid-state electronics. The common currency of these electroncis circuits was interrupted electricity and this provided a clear link to the emerging technology of communication. The combination of this data processing and logic capability with that of communication technologies have spawned a new family of 'informatics technologies'.

The development of electronics was associated with a number of very important factors. First was the growing capability to include an ever-expanding number of logic gates (ie bits) on each integrated circuit, as is shown in Figure 2.7. Secondly, largely as a result of this trend to the biannual doubling of capacity per circuit, the price of electronic components dropped continuously (see Figure 2.8) despite a general tendency in the same period towards high levels of inflation in all the major economies. Third, the advent of the microprocessor provided flexibility to the ever-more powerful circuits which were being developed (see Figure 2.9). This was a key advance since

Fig. 2.6 Binary logic as applied to introduction of office automation equipment

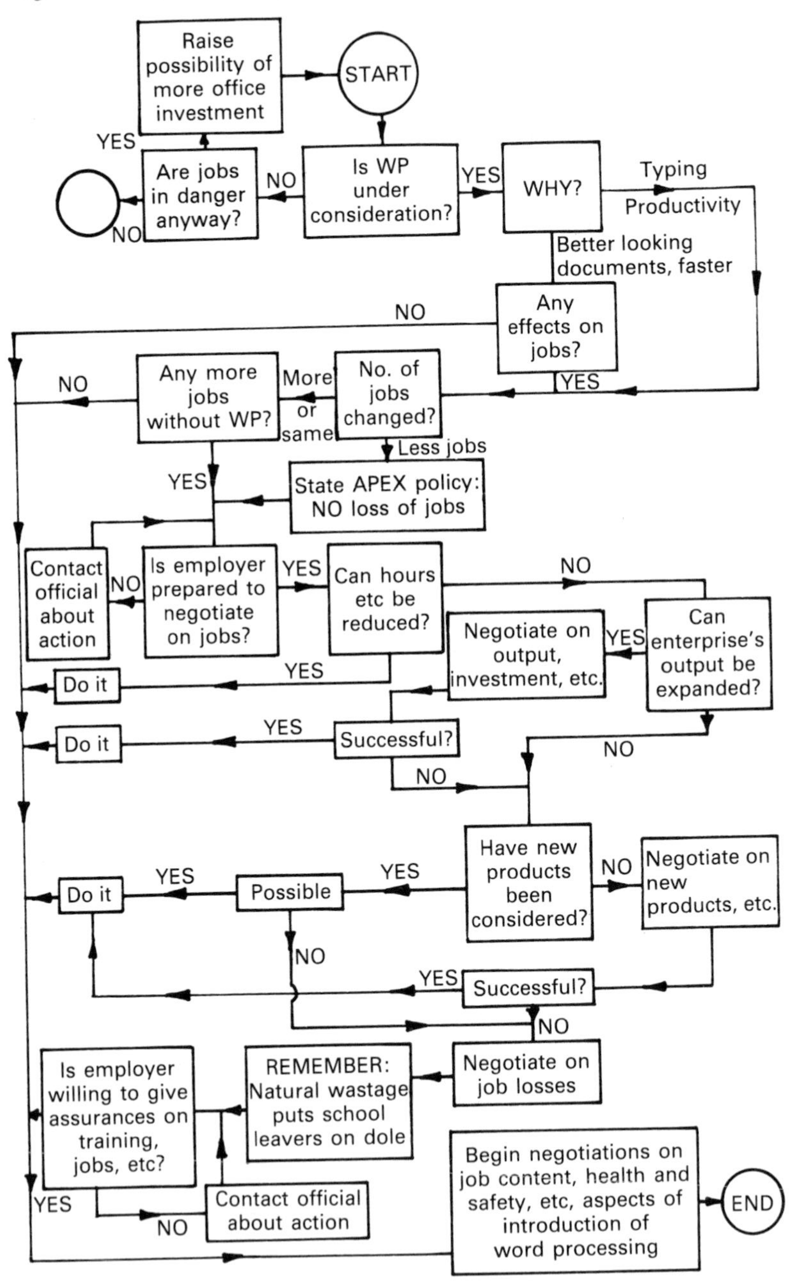

Source: Association of Professional, Executive, Clerical and Computer Staff, 1979

Fig. 2.7 Number of components per integrated circuit

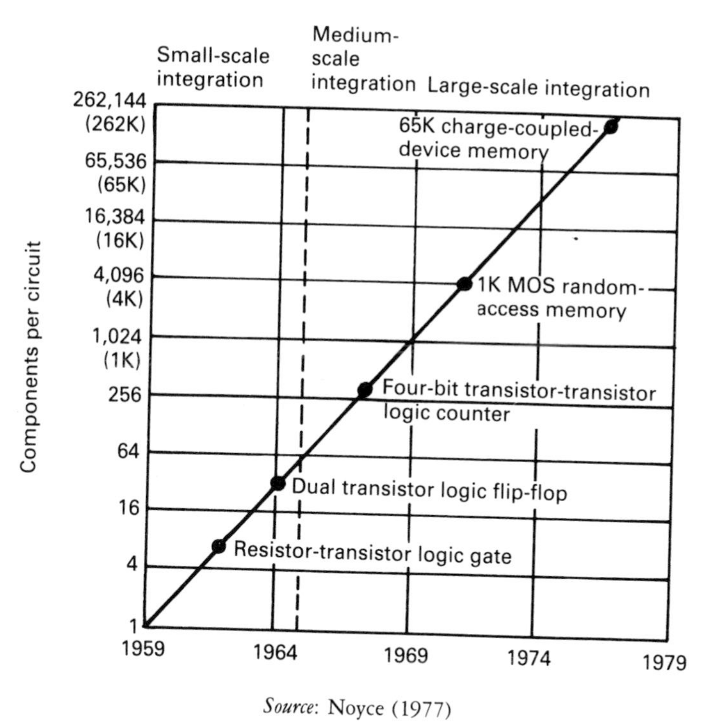

Source: Noyce (1977)

the lack of programmability made it difficult for many users to take advantage of growing complexity and reduced unit prices. And finally, there has been a consistent increase in the reliability of electronic components.

Whilst the diffusion of electronics during the upswing of the long wave was centred on the introduction of new products, it was not confined to these uses. Over the years, there was also a gradual downstream diffusion of electronics in each of the three spheres of production. The earliest use was in the design-sphere, from the mid–1950s, where the number-crunching capability of early mainframe computers allowed for the consideration of more complex design alternatives and also gave greater assurance to the evaluation of these alternatives. This initially occurred in the defence-related aerospace sector, but then rapidly spread to other civilian-oriented engineering industries. The next major steps occurred in the early 1960s when mainframe computers began to be used for processing information in large enterprises, particularly in relation to payrolls and stock-control, later diffusing to medium-sized firms. The third phase (from the mid–1960s) saw the gradual diffusion of electronics into the manufacturing sphere as firms began to take up numerical control technology.

The various elements of this electronic jigsaw are now available within most activities in each of the three spheres of production. But the rigidity of these

Fig. 2.8 Cost per bit of computer memory

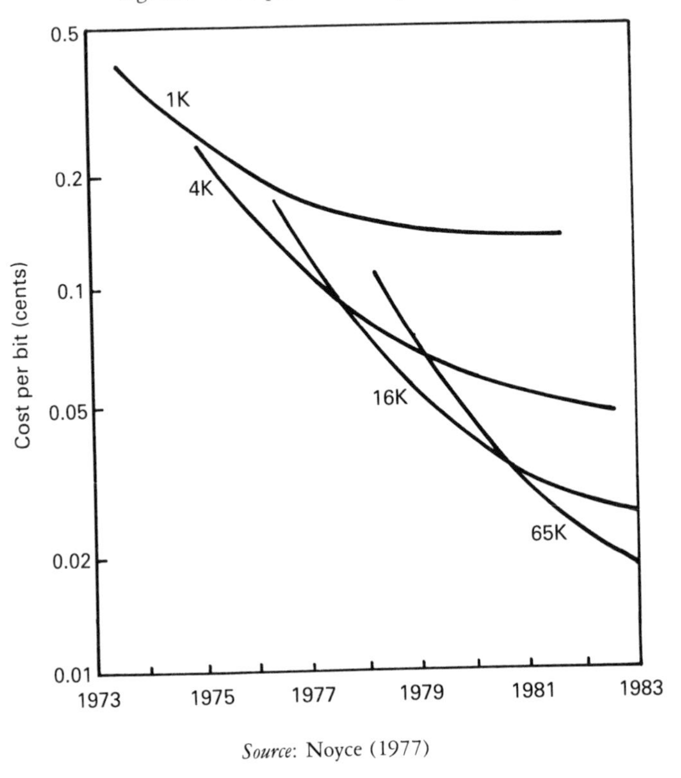

Source: Noyce (1977)

Fig. 2.9 Complexity versus commonality in electronic components

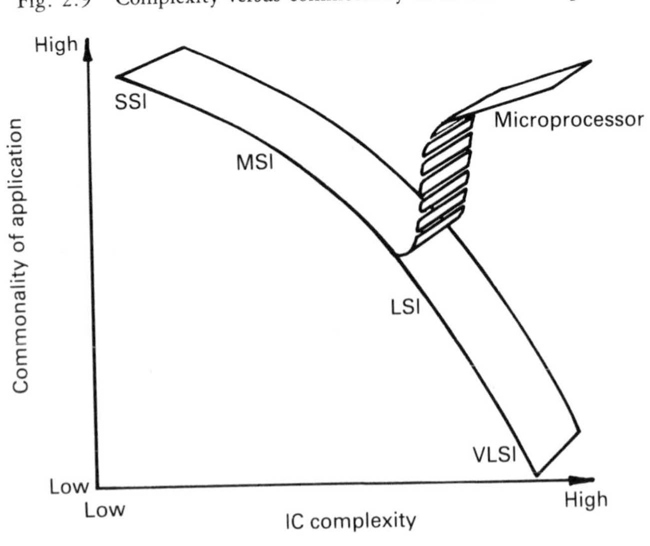

Source: Mackintosh (1978)

systems means that most of the electronics systems are confined within particular activities. For example, the development and very rapid spread of the microcomputer and then the microprocessor in the 1970s allowed for the development of intra-sphere automation systems such as computer-aided design or office-automation systems, but these were not linked to complementary electronic systems in other spheres. As we shall see in later chapters, whilst the 1960s and 1970s saw the diffusion of intra-activity and intra-sphere automation technologies, in the 1980s and 1990s the major automation advances are likely to be of an inter-sphere nature, during which the separate components of the electronic jigsaw will be joined up.

This advance in automation technology is now heavily dependent upon the downstream diffusion of electronics, notwithstanding the continued bottleneck of unsuitable sensing technologies in particular sectors, and despite the continued advances being made in materials technology. So great is this dependence on electronics that as Einzig (1957) notes, some observers have mistakenly identified automation with electronics, which as can be seen from the preceding discussion, is clearly not the case.

In the discussion which follows in Chapters 3, 4 and 5, we identify the major activities within each of the three spheres, and proceed to discuss the diffusion of intra-activity and intra-sphere automation within each sphere. Then in Chapter 6 we draw these strands together in an analysis of inter-sphere automation. Necessarily, given the importance of informatics to the progress of automation, we focus very largely in these chapters on the diffusion of electronics, but we maintain our awareness of trends in materials technology, energy technologies and sensors.

PART TWO

The Progress of Automation

In the first part of this book we defined three types of automation, namely, intra-activity, intra-sphere and inter-sphere technologies, and set these in the context of the modern industrial enterprise. Chapters 3, 4 and 5 begin by setting out the activities in each of the three spheres. Then the relevant automation technologies are described and their past diffusion charted. Finally, the benefits reaped by successful user-firms are analysed, particularly in relation to intra-activity and intra-sphere automation technologies. Part Two concludes in Chapter 6 by examining the potential for inter-sphere automation in the 1980s and 1990s.

CHAPTER THREE

Automation in the Design Sphere

THE ROLE OF DESIGN

In the previous chapter we described the historical evolution of the modern enterprise from the small, undifferentiated workshop in which only the most rudimentary specialization of work was involved to the modern, large, multi-divisional, multinational enterprise in which three spheres of production have emerged. We noted, however, that there has been great unevenness in this evolution, over time, between sectors and between countries; we also noted that we were describing a general tendency, rather than a clearly defined step-ladder of evolution. The development of the design sphere, as we shall see, reflects both the sectoral unevenness and the sporadic nature of the evolution of the modern enterprise.

In the increasing application of scientific principles to design – which underlies the evolution of a specialized design sphere – two particular strands of development are evident, both having their roots in the industrial revolution of the eighteenth century. The first of these saw the specialization of machinery manufacture, a process already remarked on by Adam Smith:

> All the improvements in machinery, however, have by no means been the invention of those who had occasion to use the machines. Many improvements have been made by the ingenuity of the makers of machines (Smith, 1950, p9).

Thus in the early period, even while the three spheres of production may not have been well differentiated within the machinery manufacturing enterprises themselves, at a wider social level the design function came to be separated to some extent from the production of commodities. And, of course, this specialization between enterprises has become an increasingly significant phenomenon, so that the development of the capital goods sector (as these machinery manufacturing enterprises are known) has come to be seen as one of the key elements of industrialization policies being pursued in the Third World.

Secondly, the specialization of design became evident within enterprises. In the early days, as the following graphic account by Adam Smith illustrates, the design process was haphazard and was not based on formalized scientific or technological knowledge.

In the first fire-engines [Smith is referring to steam-powered engines here], a boy was constantly employed to open and shut alternately the communication between the boiler and the cylinder as the piston either ascended or descended. One of these boys, who loved to play with his companions, observed that, by tying a string from the handle of the valve which opened this communication to another part of the machine, the valve would open and shut without his assistance, and leave him at liberty to divert himself with his play-fellows. One of the greatest improvements that has been made upon this machine, since it was first invented, was in this manner the discovery of a boy who wanted to save his own labour (Smith, 1950, p9).

But in the nineteenth century, the development of a specialized engineering and design profession became more explicit, involving 'the conscious application of science, instead of the rule of thumb' (Marx, 1976, p386). However, the crucial factor which led to the differentiation of design (with research and development) into a specialized sphere of production was the introduction of flow processes in the chemical industry in the latter part of the nineteenth century. The following quote from Freeman's seminal study on industrial innovation makes this point clearly:

This change [ie the introduction of professionalised design within industry itself] has affected especially the design of new products, but the new science-related technologies also affect the way in which improvements and changes are made in production . . . in the older industries these could be made predominantly 'at the bench' by direct participants in the production process. The subdivision of mechanical processes did not remove this possibility. Indeed, as both Adam Smith and Marx noted, the workers themselves were often responsible for inventions leading to further subdivision. But the introduction of flow processes in the chemical industry [in the second half of the nineteenth century] and of electronic control and automation in other branches of industry [in the twentieth century] mean that improvements and changes now depend increasingly on an understanding of the process as a whole, which usually involves some grasp of theoretical scientific principles. *It also means that experiments usually have to be made 'off-line' by production engineers or operatives. All this has accentuated the relative importance of the specialised R and D group or Engineering or Technical service department and diminished the relative importance of the 'ingenious mechanic'* (emphasis added) (Freeman, 1974, p30).

Thus, we see the evolution of design (incorporating R & D and technical services) and its separation from the manufacture and overall coordination of the enterprise. Already in the late nineteenth century engineering had not only become a professionalized activity with its own set of reference groups, but within the profession there was an increasing separation between those engaged in 'trouble shooting' activities on the shop floor and those involved in design, R & D and technical services. The first group worked in the manufacturing sphere, whilst the latter were hived off into design.

ACTIVITIES IN THE DESIGN SPHERE

In order to understand the role played by design in the enterprise, it is helpful to view the production of commodities (particularly durable ones) and services in the light of systems. Here we might lean on insights gleaned from the systems engineering literature.

> Successful planning and design of large complex systems requires the 'systems approach'. The systems approach recognises that factoring out a part of a problem by neglecting the interactions among subsystems and elements increases significantly the probability that a solution to the design problem will not be found; it requires that the boundaries of the system be extended outward as far as is required to determine which interrelationships are significant to the design problem.
> A system to be useful must satisfy a need. However, designing a system to just meet the need is not usually sufficient. With few exceptions, the system must be able to meet the need over a specified period of time in order to justify the investment in time, money, and effort. Thus one must consider a system in a dynamic sense – the life cycle or so-called 'cradle to grave' viewpoint. The system life cycle may be said to originate in the perception of a need and to terminate when the system becomes obsolete (Kline and Lifson, 1968, pp12–15).

Thus design must be seen as a subset of a variety of interrelated activities which include production, installation, operation, maintenance and modification. Unless we recognize both this temporal dimension to product development and utilization, and the interrelationship between these different subsets of activities, we will be unable fully to appreciate the coming discussions on the potential of inter-sphere automation technologies (Chapter 6).
In general terms, as can be seen from Figure 3.1, design comprises six major sets of activities.

Specification

The product or design problem first has to be specified. This can be done by the enterprise if it has some view of the demand for the product. It may be very tightly defined (eg in a tender document) or loosely based upon a perceived 'hole' in the market. Clearly this product specification involves an interaction with both the 'external world', either through the customer's specification or an evaluation of the market potential for the product, and the 'internal world' in which through discussions with the coordination sphere of the enterprise the new product under development is situated within the broader strategies chosen by upper management for the firm's future (see Chapter 5).

Basic design

Once the product has been defined it is necessary to proceed with a basic design. This activity generally involves a high 'science' input (although the degree clearly varies between sectors, firms and products) in which the essential design principles are explored. For example, it is unlikely, given the knowledge avail-

Fig. 3.1 Activities in the design sphere

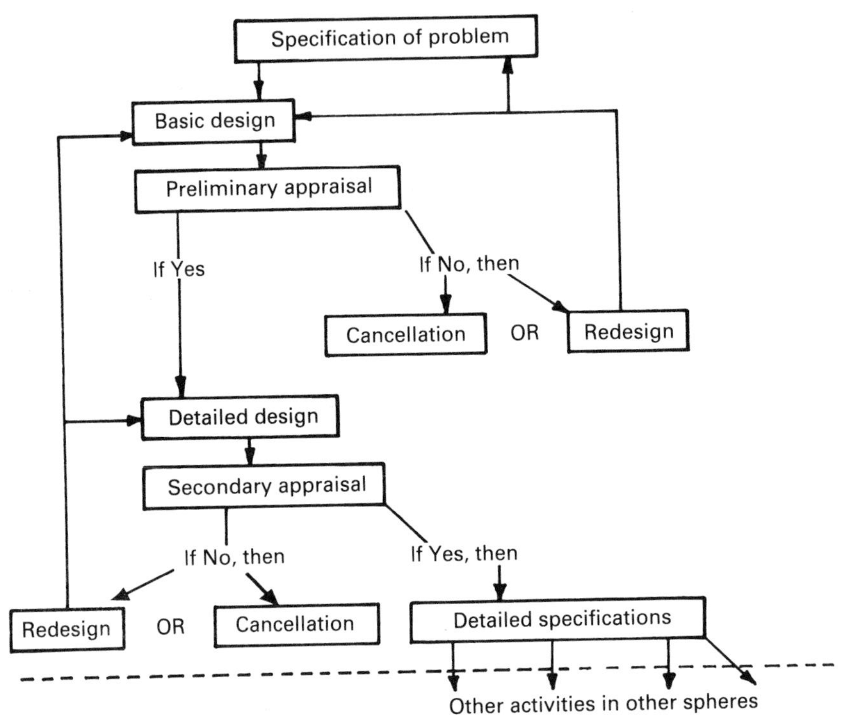

able to most enterprises, that the basic design of an engine would explore the use of water as a fuel source; but, by contrast, it is likely that a basic design for a turbocharged diesel internal combustion engine may well be explored. Here the design activity will draw upon knowledge which is both internal to the firm as well as from the 'shelf of knowledge' in the external world (to the extent that it is not 'owned' by other enterprises).

Preliminary appraisal

Once the basic design is complete it is necessary to evaluate it to determine whether to proceed with the next activity, namely detailed design. In this case, through interaction with both the coordination sphere and (to a lesser extent) the manufacturing sphere, the attractiveness of the product is examined. The sorts of questions to be answered in this phase are: is its manufacture feasible within the constraints faced by the enterprise; are material inputs available at a reasonable cost; does the market exist given the likely costs of production; does the enterprise have the necessary skills and capacity required to ensure timely production; what degree of risk is involved; and, perhaps most importantly, what are competing firms doing? Only once these, and other, basic questions are answered in the affirmative and with a reasonable degree of confidence, will it be possible to proceed with detailed design activities. If the answer is negative the basic design is either respecified or the intended product jettisoned.

Detailed design

Having passed the preliminary appraisal, the design is then pursued at a more detailed level, involving the numerous components of the product. For example, a basic design decision may have been made to go ahead with the production of a new fuel-efficient motor car. In the detailed design phase, the precise dimensions of the body panels, the seating arrangements and so on will be specified. It is clear from this that the detailed design activity differs from basic design not only with respect to the degree of detail, but also in relation to the type of knowledge which is involved. In general basic design involves greater attention to underlying scientific principles, much of which is to be found outside of the enterprise, whereas detailed design is more technologically oriented and relates more closely to other spheres of production within the enterprise and to historic experience within the firm in relation to previous products.

Secondary appraisal

During this activity the detailed designs are subject to greater scrutiny. But in this case the appraisal is concerned more with product and process optimization than with the basic yes/no decision. Only if a major design flaw is discovered or if market conditions change in a significant way is it likely that the secondary appraisal will question the decisions reached in the primary appraisal activity. Then if the secondary appraisal is affirmative the design process moves to its final activity (detailed specifications); if negative the product is subject to changes in the basic or detailed design, or (in isolated cases) is jettisoned altogether.

Detailed specifications

The final activity in the design sphere is the preparation of detailed specifications which are then passed to the other two spheres of production. Manufacturing instructions are passed on to the manufacture sphere, generally in the form of detailed drawings made by draughtspersons. At the same time the coordination sphere is provided with parts lists, bills of materials, product descriptions and specifications of machinery and inputs so that it can coordinate these in an effective manner.

TECHNOLOGIES FOR AUTOMATING DESIGN

The history of CAD

As can be readily seen from Figure 3.1, there are a limited number of technologies involved in this sphere. Basically these are utilized for calculation of design parameters and for presentation of information (usually documents and drawings). The traditional pre-electronic equipment used in this sector was slide rules, drawing boards and typewriters; not surprisingly therefore the 'capital equipment' backing each worker in the design sphere was reckoned to be very low, often less than one-tenth that in the manufacturing sphere. Even in the early 1970s there were many design offices in which the equipment was limited to drawing boards, slide-rules, writing implements and typewriters.

However, as technology became increasingly science-based these tools proved to be inadequate. The first major technological developments were stimulated by the need to ensure greater accuracy in the design of shell-trajectories in the second world war, which stimulated the early development of the (pre-electronic) computer. And as computer technology began to mature in the 1960s it was increasingly applied in the design sphere (especially in military related projects). Right through to the early 1970s this involved the use of cumbersome mainframe computers which whilst offering increasing power and speed for lower prices, required an organizational system in which design alternatives had to be explored on a batch basis, a procedure which constrained the usefulness and diffusion of computers in the design sphere.

The development of the minicomputer in the early 1970s provided a solution because it allowed not only for interactive use (ie by obtaining an 'immediate' answer to design changes), but also for decentralized use by different divisions in the design sphere. However, the availability of these computers to undertake design calculations was in itself not a sufficient factor, because traditionally designers had come to work and think through the use of drawings. The first embryonic graphics technologies had begun to emerge in the 1960s to fit the needs of the electronics industry in designing circuits (called 'digitizing tablets'). They converted drawings into numerical coordinates which were suitable for treatment by digital logic (see Chapter 2). Subsequently these numerical coordinates were used to produce pictures on a television-type screen, hence giving rise to computer-graphics.

Thus it was the marriage of these three sets of electronics-based technologies – that is minicomputers, digitizing boards and visual display units – that led to the development of the major automation technology used in the design sphere, namely computer-aided design (CAD). By CAD the industry generally refers to interactive computer-graphics rather than to the non-graphic, batch-orientated use of computers in design which was a characteristic of the 1960s and 1970. As we shall see, the industrial origins lay in the defence/aerospace sector, after which it made the transition to the electronic and automobile sectors and thereafter spread (at an 'explosive' rate) to other sectors in the late 1970s.

Fig. 3.2 Typical CAD configuration

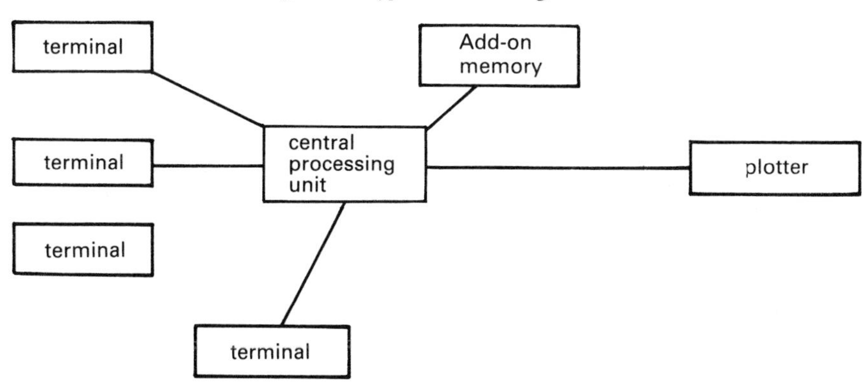

In essence a CAD system involves four different sets of hardware, as illustrated in Figure 3.2. Of primary importance is the computer, the central processing unit which controls the sequence of processing, undertakes the arithmetic tasks underlying digital logic and stores necessary information. Data then has to be put into this computer, and here there are two major types of workstation. The first is the digitizer board which converts designs to numerical coordinates (ie binary code), and the second is the visual disply unit (ie a television-like screen) which allows the designer to see the design and to interactively proceed with or amend it. Once the design has been completed it has to be communicated: in paper-based systems this may take the form of drawings produced on various types of plotting devices; in automated systems, the design is transmitted in the form of numerical coordinates, often on paper tape. The final set of hardware is add-on memory, since with current technology the available computers do not possess sufficient core memory of their own. Here there are alternatives (basically disc and tape), each of which has its advantages and disadvantages and are often used together in the same CAD system.

On its own this hardware, all of which is readily available from a fairly large number of suppliers, is useless since it requires a series of operating instructions – that is, software – to operate. There are two different elements of software which are necessary for draughting, with an additional capability for design. The first of these is the operating system which specifies the way in which the computer executes its tasks – all computers require an operating system. But to produce drawings it is necessary to also have basic graphics software, which is the ability to construct lines, circles, arcs, rectangles, mirror-images, etc on the screen and subsequently on to paper or paper tape. The software required for graphics is relatively simple and is now widely available. But the use of the CAD system for design requires specific and more complex applications software. The earliest applications programs were developed for the electronics sector. Subsequently they began to meet the needs of the mechanical engineering sector and the push now appears to be occurring in the architectural, civil and structural engineering sectors. There already exist a wide variety of different applications programs, the development of which (as we shall see) are the heart of the activities of CAD suppliers. These range from by now fairly standard routines (eg the calculation of optimum routes for linking the electronic components on a printed circuit board), to those which are actively under development (eg solid modelling). Some of these applications programs are not heavy users of data (eg autorouting), whereas others (such as finite elements stress analysis) require powerful mainframe computers for execution. From the operator's point of view the primary method of using these applications programs is via 'menus'. These are generally removable sensitized tablets which contain specific routines relevant to particular applications programs and which enable rapid use. An analogy to changing menus would be the change from a statistical calculator (with buttons for regressions, square roots, etc) to a scientific calculator (with buttons for trigonometric functions). Characteristically, an efficiently run CAD system will have very many menus, each designed to fit a particular set of applications.

The distinction between these variants of software is of critical importance to understanding issues such as the dynamics of market structure, the benefits arising from the use of CAD, the skills it requires and their associated learning curves. Currently there are a fairly large number of firms supplying both packages of CAD systems and/or individual hardware and software components of such a system. Moreover, most of these suppliers provide a wider range of options. But the most significant of these arises from the type of computer which is used and this largely defines three segments of the CAD market. These three alternatives are as follows.

Microprocessor-driven dedicated terminals are small and not very powerful. basically they are suitable as pure draughting aids – a sort of draughtsperson's word processor – although some are also able to undertake elementary processing programmes such as laying out the electrical circuits on a printed circuit board.

Minicomputer-driven systems are more powerful and more flexible. These form the basic processing capability for all of the existing turnkey systems. Their strength relative to the small dedicated systems is that they are powerful enough to be able to undertake a large number of applications programs as well as to function as a draughting tool; each minicomputer is also able to drive between three and eight terminals, depending upon the particular supplier's software and the use made of it by the user.

The power of *mainframe* computers provides two major advantages to users. The first is that these systems are powerful enough to undertake the more taxing requirements of particular software applications (eg finite element modelling in mechanical engineering) as well as to also process databases (eg parts lists, payrolls, etc) for which minicomputers are not suitable. And second, the power of the mainframe allows large users (or those using them on a time-sharing basis) to reap economies of scale in unit terminal costs.

THE DIFFUSION OF DESIGN-AUTOMATION TECHNOLOGY

Running parallel to this development of CAD technology has been an explosive growth in demand for design-automation equipment which has transformed a small, esoteric, high-technology industry into a major and pervasive tool for manufacturing industry in general. As can be seen from Figure 3.3 the industry has grown at a phenomenal pace, particularly in the 1976–81 period during which the technology began to diffuse through to the mechanical engineering sector. Thus from global sales of around $70m in 1976, the aggregate turnover of the ten or so major vendors of turnkey CAD systems rose to over $1 billion by 1980; this represented a compound annual growth rate of 70 per cent pa over the five years, rising to 85 per cent pa in the 1978–80 period. And even in the 1981–2 period when almost every sector of American industry (including that of electronic components) was being severely affected by the recession, the industry growth rate was over 40 per cent. Figure 3.3 also indicates the size of the CAD industry if it had grown at the same rate as DEC (the most successful of the minicomputer firms) and IBM (the giant in the mainframe computer sector) – clearly the growth of demand for design-automation equipment has outstripped that for electronic equipment in general.

Of course many new industries show remarkable growth rates in their early years and this is not too difficult when the base year aggregate figures are small. But the CAD industry turnover figures are no longer in this 'small league'. For example, the 1981 aggregate turnover of over $1 billion excludes the sales of the business-graphics (to the coordination sphere of production) and animation sectors which were together in excess of $300m; but even if the base figure is taken as $1 billion, and the industry grows at a conservative projection of 40 per cent pa the likely 1984 (global) turnover of over $4 billion compares favourably with projected sales in the US colour television sector of around the same figure.

This tranformation of the supplying industry is reflected in the sectors using the technology. In the 1950s CAD technology was developed primarily for the US early-warning, nuclear defence industry. Diffusion to the aircraft and automobile industries began in the late 1960s, but it was the requirements of the electronics industry which led to the establishment of the new independent CAD-supplying firms such as Computervision, Applicon and Calma who now dominate the industry. Thus in the first two-thirds of the 1970s, the basic software which the industry now leans on was developed primarily to meet the needs of the electronics industry. Having matured, however, the CAD-supplying industry soon learned that there existed an enormous untapped demand for design-automation equipment in the mechanical engineering industry. Penetrating this accounted for the phenomenal growth in the 1976–81 period.

As for the 1980s the supplying industry anticipates three major areas of growth. The first is in CAD/CAM (Computer-aided design/computer-aided manufacture), which represents one element of inter-sphere automation as defined in Chapter 2. The second is the use of business graphics in the coor-

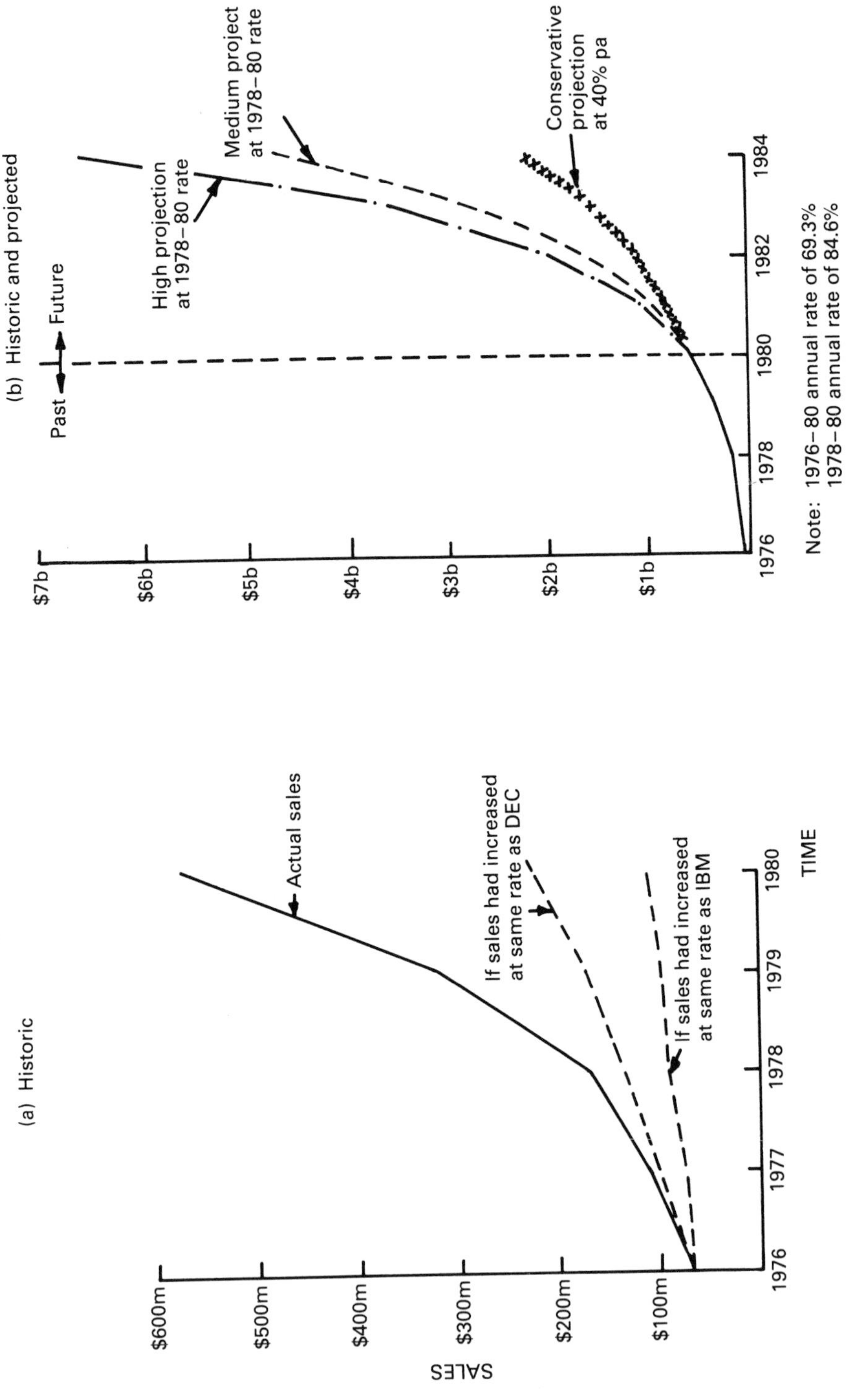

Fig. 3.3 Sales of US turnkey vendors – past and projected

(a) Historic

SALES

$600m
$500m
$400m
$300m
$200m
$100m

1976
1977
1978
1979
1980

TIME

Actual sales

If sales had increased at same rate as DEC

If sales had increased at same rate as IBM

(b) Historic and projected

Past ← → Future

$7b
$6b
$5b
$4b
$3b
$2b
$1b

1976
1978
1980
1982
1984

High projection at 1978–80 rate

Medium project at 1978–80 rate

Conservative projection at 40% pa

Note: 1976–80 annual rate of 69.3%
1978–80 annual rate of 84.6%

dination sphere (see Chapter 5) where one study predicts that demand will grow at twice the rate of that seen in other CAD sectors (Kaplinsky, 1982b). And, third, many CAD suppliers are specifically orienting themselves to the vast and relatively untapped market in the architectural and civil engineering (ACE) sectors. By 1990, therefore, it is widely anticipated that the design-automation industry's sales will exceed $14 billion.

Finally, it is worth noting that despite the remarkable sales growth rate and the aggregate size of the sector, most observers estimate that although penetration in the electronics industry is fairly high, only around 5 per cent of the potential US CAD/CAM market in mechanical engineering has yet been tapped. (The US market, whilst showing the greatest degree of penetration is now growing less rapidly than that of Japan and Europe). Moreover, the business graphics and ACE sectors have barely been tapped and present an enormous reservoir of potential demand.

THE BENEFITS PROVIDED TO USER-FIRMS

Clearly these heady historic and anticipated growth rates are not merely based upon a perception of hoped-for benefits. The technology must offer some rather dramatic benefits for the market to grow so rapidly particularly during a period of emerging depression. It is worth detailing some of these gains as an indication of the sorts of benefits being offered by the new, electronics-based automation equipment discussed in this book. As we shall see, it corroborates three of the main points discussed in the first chapter. First, that in an era of supercompetition, automation has a critical role to play in enabling capital to face the advances made by competitors. Second, by changing the nature of work – a discussion which will be pursued in greater depth in Chapter 8 – it provides capital with a technique for coping with independently minded labour. And, third, the development of the CAD industry illustrates the transition in the current long wave cycle from the upswing of the 1950s and 1960s – where the heartland, electronics technology was predominantly used for the introduction of new products such as integrated circuits – to the rationalizing downswing where the heartland technology is predominantly used in established industries to cope with growing competitive pressures.

But before we discuss the benefits provided by design automation technology – which we shall do in relation to two of the three types of automation defined in Chapter 2, namely intra-activity and intra-sphere (inter-sphere automation will be discussed in Chapter 6) – it is important to bear in mind that we refer here to the benefits provided to capital, and not labour. We do so because, as pointed out in Chapter 1, automation technology is diffusing rapidly precisely because it offers an escape route to individual units of capital from destruction due to supercompetitive markets. Consequently it is important to specify the sorts of benefits which capital obtains, although, for capital as a whole it may not provide the same escape route (see Chapter 10) since, amongst other things,

CAD technology and other automation equipment displaces labour and hence further reduces demand in a period of chronic overcapacity. The impact on labour of using these automation technologies is a separate issue, and is treated later in Chapter 8.

Intra-activity automation – product benefits

Referring back to the various activities outlined in Figure 3.1, it is evident that design-automation technology has a role to play in all but the first activity, that is in basic design, preliminary appraisal, detailed design, secondary appraisal and the drawing-up of detailed specifications; it is, however, relatively useless in the process of specifying the product to be made. The primary benefits obtained by most CAD users lie in the activities of basic and detailed design, and in drawing detailed specifications. These reflect improvements in both product and process.

In the area of product benefits, we can distinguish between three subsets of advantage which have been realized by user-firms. For some CAD is an *essential tool*: there are a variety of products which could not be produced without CAD, such as VLSI integrated circuits, nuclear power plants, military radar and aeroplanes. While in many of these cases (with the possible exception of some electronics applications, where visual interaction is essential) designs could be produced with non-graphics-based computer-aided design, the absence of an on-line, interactive relationship (ie via graphics terminals) between designer and product would almost certainly make the final designs suboptimal and uncompetitive in the market place.

In an increasing number of sectors CAD is an *optimizing tool*: the differentiation and optimization of products is essential as competitive pressures hot up during the supercompetitive downswing of the long wave cycle. CAD has an essential role to play in allowing firms to modify and optimize their final products more effectively. This is particularly true in the 'capital goods' sector, loosely defined to cover large projects where the final product tends to be individually tailored to the customer's needs and where design costs are a large proportion of total costs. An example drawn from the aircraft sector illustrates this graphically. In civil aircraft:

> R&D costs are shown to account for about half of the total launch costs. Design costs generally constitute around 25 per cent of the R&D costs. Hence, 'design' accounts for about 12½ per cent of the launch costs, and this is about 1 per cent *only* of the selling price.
>
> However, during the spending of this 1 per cent the cost effectiveness of the product is essentially defined. Design exerts a controlling influence on all elements of selling price and subsequent in-service costs, as well as having the responsibility for all aspects of performance achievement (Jacobs, 1980, p178).

Figure 3.4 drawn from British Aerospace studies of cost control, provides further evidence of these benefits. In it, the life cycle of an aircraft is divided into five phases, namely conception, validation, development, production and

Fig. 3.4 Life cycle cost – money committed vs money spent (Civil aircraft)

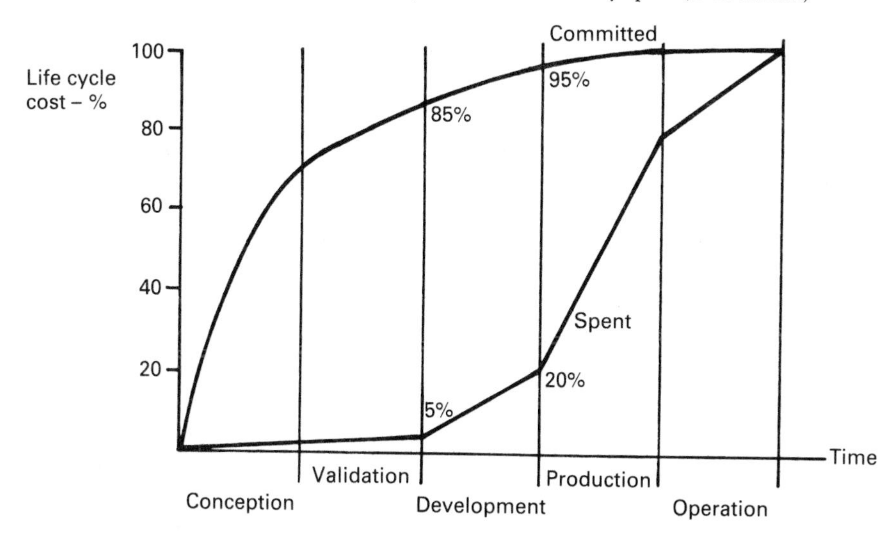

operation. The first two-thirds of this life cycle is accounted for by design and in this design cycle, while only 5 per cent of the total life cycle costs of the product are incurred, fully 85 per cent of the total lifetime operating costs of the aircraft are determined. In this context, optimizing the final product is of critical importance, far more so than any savings in labour costs which might arise from the automation of design and draughting with CAD equipment.

Civil aircraft may appear to be an unusually clear case of the advantages derived from the optimization of product, but is is not unique. For example, a major sphere of competition in the automobile sector is associated with the need to make cars more fuel efficient. This involves an optimization of engine design, weight minization and a reduction in the drag coefficient (Senker, 1980) – all of these problems are currently being tackled with the assistance of CAD. Another rapidly emerging example concerns the layout of retailing premises and fast-food restaurants: one of the major chains of fast-food vendors explores alternatives and optimizes space utilization through CAD. And a final example points to the very substantial benefits which can arise from the use of CAD and which do not arise from productivity improvements in draughting. Increasingly, the boundaries between oilfields are defined by computarized systems, and this is for a very important reason. In a typical North Sea oilfield, misplacing a boundary by one metre can involve a loss of $60m in oil, consequently the oil companies insist upon the use of computers in defining their oilfield boundaries.

The third main advantage is *product lead-time*. While the cost of final products and their performance are critical areas of market competition, the timeliness with which products are launched is very often of even more crucial significance to particular firms. CAD plays a very important role in this regard through the reduction in lead-times which it permits. This benefit is really the other side of the same coin which enhances the productivity (per labour hour) of design offices.

While numerous examples could be given to illustrate the benefit derived from the shorter lead-times which CAD permits, two are particularly pointed. The first concerns a British firm which had been through various stages of asset-stripping ownership. Close to bankruptcy, one of their design engineers came up with a radical redesign of its major product which had to be launched within one year or the firm would go out of business. On the basis of archival records it would have required 12 person years of draughting, but not only was there a shortage of draughtspersons in the area, but much of the design was sequential. (So the process would have taken at least 18 months which was beyond the one year survival limit given to the firm by its bankers.) A two-terminal CAD system was procured at short notice and three designers produced 8,000 drawings in the first year, compared to the archival records of 400 drawings per person per year in the old manual office. In the second example of a British machinery manufacturer making effective use of CAD for modular plant designs, a designer was telephoned at 6 am on a Saturday morning by a salesman about to go to Mexico the same afternoon, who gave details of a tender for which the firm was able to compete. By noon the salesman was given a complete tender document including drawings, prices, bills of materials and estimated duration of construction, and the order was obtained. In both these cases – which are merely examples of a commonly observed process – without the rapid response enabled by CAD, the firms would have failed to maintain their market presence.

Intra-activity automation – process benefits

The major benefit in process arises from the reduction of costs, specifically those of draughting labour. There are a variety of factors at work in determining the productivity which the technology actually operates: this is not just the speed (and precision) at which the lines are physically drawn on the screen, but includes the routine background calculations which lie behind the drawing. For example, changing the specifications of a biscuit-machine from one size to another, as occurred in the case of one user, requires the re-scaling of all components of the design: manually, this involves endless hours of routine calculation whereas on the CAD system used, the scaling is done automatically and almost instataneously. A second and related factor is that it depends upon the type of drawing involved. In primary design drawings, the productivity of the CAD system is seldom higher than manual systems, but on modifications the productivity gains were substantially higher, frequently over 20:1 and in one case computed at 100:1. And thirdly, many of the gains from CAD depend upon the system with which the design office is organized. This is particularly important when the product is of a modular nature and particular parts can be stored in the memory of the CAD system; instead of manually redrawing a particular hopper or conveyor-belt in a plant design, the CAD operator can instantaneously recall it (or a variety of alternatives) from memory and append it to the layout under design.

Consequently there is an enormous variation in the productivity of CAD systems. Nevertheless, a number of enterprises have actually computed the overall average productivity of their CAD systems and the results are shown in Table 3.1 below. In most cases these productivity ratios were realized despite the fact that CAD was being used primarily as a design-tool, rather than as an automated draughting-aid. (CAD/manual productivity ratios in design are almost always lower than those in draughting.) On average, most of these users were realizing productivity gains of over 3:1 over manual systems. Consequently one large-scale firm estimated that it would displace aroud two thirds of its draughtspersons before 1985.

In addition to reducing costs through the saving of labour time, design-automation offers another important attribute in improving the quality of drawings produced. One top-quality master copy can be maintained, a major advantage when compared to the more haphazard and non-automated systems where a number of grubby copies – each amended in different ways – tend to float around different activities and in different spheres of the enterprise.

Table 3.1 Productivity of CAD systems

Sector of activity	*Location*	*Primary use of CAD*	*Average productivity ratio*	*Range of PR between different types of drawings*
Integrated circuits	US	Design	2:1 after 6 months	NI
Automobile components	UK	Design	3:1 after 12 months	NI
Plant design	UK	Draughting	3:1	1:1–20:1
Process plant	UK	Design	NI	1:1–50:1
Electric motors	UK	Draughting	6.6:1	NI
Printing machinery	UK	Design/draughting	>2:1	NI
Architecture	UK	Design	3.5:1	NI
Automobiles	UK	Design	3:1	NI
Computers – pcb's	UK	Design/draughting	>5:1	NI
Process plant	UK	Design/draughting	4:1	NI
Petroleum exploration	UK	Design	2:1	NI
Automobiles	UK	Design/draughting	2.78:1 after 6 months	NI
Aircraft	US	Design	2.5:1 in 1979 3.32:1 in 1980	NI NI
Instruments – pcbs	UK	Design/draughting	>3:1	NI
Public utility	US	Draughting	>3:1	

NI = No information

Source: Kaplinsky (1982b).

Intra-sphere automation

Although many users of design-automation equipment are still in the learning phase – particularly those in the non-electronic and non-aerospace sectors – some firms are already reaping the gains of intra-activity automation. For example, Boeing Aircraft now largely appraise their detailed and basic designs through their CAD systems rather than by testing mock-ups in wind tunnels. Another example can be drawn from a new system being introduced by Daimler-Benz in Germany:

A motion database will simulate truck and automobile dynamics while the visual system will depict road conditions (ice, snow, dry); roadway settings (urban streets, suburban roads, expressways, and hilly countryside paths); different conditions of visibility (day, night, fog, twilight); and even animated traffic and stop light conditions (cross traffic, parallel traffic). When simulating night-time driving, headlights will illuminate scenery ahead . . .
Using the simulator, scientists and engineers can test the study the effect of skidding, collisions, ordinary driving manoeuvres, braking and the like. In operation, a driver sits in the cab equipped with actual instruments and control panels, responding to various conditions that appear in the visual display. Scenery will be projected on a curved screen providing simulated windshield and side-window views in a 180° field of visibility (*The S. Klein Newsletter on Computer Graphics,* vol 4, no 6, 1982).

Here we see an example of intra-sphere automation in the design sphere which represents links between four sets of activities: basic and detailed design and primary and secondary evaluation. As described, it falls short of complete intra-sphere automation in that the production of detailed specifications is not directly linked to this design/appraisal system. However, such a link-up is not problematic and many CAD users have installed systems in which the detailed specification activity (ie draughting) is directly linked to design.

This story of intra-activity automation is not confined to the automobile and aerospace industries, but is widespread amongst all users. What is more problematical, as we shall see in the discussion of inter-sphere automation in Chapter 6, is the linking up of these design automation technologies with other automation technologies in the manufacture and coordination spheres of production. But the availability and viability of intra-sphere, inter-activity automation in design – an unknown entity as recently as the late 1960s – is now beyond question. Moreover, as we discussed in Chapter 2, the building blocks of these new automation technologies are predominantly made up of electronic components.

CHAPTER FOUR

Automation in the Manufacturing Sphere

In this chapter we will focus on the emergence of intra-activity and intra-sphere automation technologies within the sphere of manufacture. In order to do so we have first to identify the basic activities in this sector. But here we face some difficulties, since unlike design and coordination where the basic activities vary little between different sectors, there are important sectoral differences which determine the nature of the activities involved in manufacture. Therefore, before we proceed to a discussion of activities in this sphere of production, it is first necessary to distinguish briefly between three sets of factors affecting the diffusion of automation technologies. These are the differences between dimensional and discrete-product industries, the scale of production involved, and the distinction between assembly and forming processes.

THE NATURE OF MANUFACTURING

Process and non-process sectors

When confronted with the distinction between process and non-process sectors, we are intuitively drawn to what appears to be a clear defining difference. Process industries are generally thought of as involving a continuous flow of production whereas by contrast, in the non-process sectors, production takes place in a discontinuous batch form. Usually we tend to think of chemical industries as involving some form of continuous process reaction, whereas manufacturing industries involve production on a non-continuous batch basis.

Yet the distinction between process and non-process industries is not as clear as it might seem at first glance. For example, many chemical industries which we might readily consider as process industries given the continuous nature of the central process itself, in fact operate on a batch basis, especially when production occurs in a multi-product plant. On the other hand in some sectors (such as modern plant bakeries), despite the inherent batch nature of production, the plant appears for all intents and purposes to be operating on a continuous basis. Thus Woodward (1965) in her study of 203 South Essex manufacturing firms in the late 1950s made this observation on the organization of batch-production firms:

> . . . the tendency was to move towards a newer and more complex system: from unit and small batch to continuous-flow and process production . . . continuous-flow production not only of liquids, gases and crystalline substances, but also of solid shapes will probably predominate in time. Indeed, it was interesting to find that even at the present time the large batch and mass production methods, normally regarded as typical of modern industry, were in operation in only one in three of the firms studied. Continuous-flow production methods were being applied in an increasingly wide range of products ; the canning and packaging of food for example (p 47).

Because of this potential confusion between continuous and batch production, in process and non-process industries, Woodward characterized the essential difference as occurring in relation to products, where she distinguished dimensional products (ie those measured by weight, capacity or volume) from integral products (ie produced as single units often called discrete-parts production). Thereafter, as we can see from Figure 4.1, she gives subsidiary significance to the contrast between batch and continuous production lines.

Scale of production

In both the cases of dimensional and integral products there will inevitably be a wide range in the scale of output. This range can be classified into a variety of subgroups, the precise classification used depending upon the particular analysis being pursued. Thus production engineering theory (which is ultimately concerned with the technology of manufacture) distinguishes between jobbing (ie single unit), batch and mass production, whilst Woodward (whose interest covered managerial organization) argued that small batch production is more usefully included in a single group together with unit production (see Figure 4.1), a view also held by Amber and Amber (1964).

From the perspective of automation, the precise categorization adopted is not terribly important. Rather, the important factor to grasp is that the scale of operation has an important bearing on the nature of automation technologies which are used. For example, as we shall see, automation has hitherto only really been feasible in relation to very large scales of output, but a major significance of emerging cybernetic-type automation systems is that their greater flexibility has begun to lower the scale at which automation technologies are competitive.

Bearing the importance of scale in mind, it is interesting to note the great preponderance of unit and small production enterprise in modern economies, particularly in the metalworking sectors. In the USA, for example, over 75 per cent of all metalworking firms are engaged in small batch production (Lund et al, 1977), and between 50 and 75 per cent of the dollar value of all 'durable' manufactured goods are batch produced (Ayres and Miller, 1983). More detailed information exists for the UK, from which we can see how the average scale of production varies by sectors. Around 50 per cent of all production in these sectors occurs in job-lots of less than 100, and 80 per cent in batches of

Fig. 4.1 Scale in production

A. Integral products

Woodward's classification	*Nature of activity*	*Production engineering classification*
Unit and small batch production	Production of units to customer requirements Production of prototypes Fabrication of large equipment in stages	Jobbing ('Unit production')
	Production of small batches to customers' orders	Batch
Large batch and mass production	Production of large batches	Batch
	Mass production	Mass

B. Dimensional products

Process production	Intermittent production in multi-product plant	Batch
	Continuous flow production in single product plant	Mass

Source: Adapted from Woodward (1965), p. 39.

Table 4.1 Relationship between industrial groups and distribution of work according to batch quantity

Industrial group	*1–5*	*6–20*	*21–100*	*101–200*	*201–500*	*501–1000*	*1000*	*Total percentage*	*Median value*
Office machinery and automotive equipment	0	0	4	1	14	10	71	100	1200
Aircraft controls and instruments	4	42	46	3	2	0	3	100	23
Miscellaneous mix of products (light)	1	3	17	14	28	15	22	100	330
Reciprocating machinery (medium)	2	1	31	27	30	7	2	100	150
Miscellaneous mix of products (medium)	5	14	36	18	16	7	4	100	86
Rotating machinery	41	22	32	0	5	—	—	100	9
Machine tools	28	31	35	3	2	0	1	100	15
Reciprocating machinery (heavy)	17	33	21	12	13	1	3	100	20
Miscellaneous mix of products (heavy)	28	37	31	1	2	1	0	100	13
Combined percentage	8	16	29	12	15	6	14	100	90

Source: Bell (1972).

less than 500 (Table 4.1). Mass production is particularly prevalent in the automobile and office machinery sectors, and the 'heavier' the industry, the greater the tendency towards small batch production.

Assembly and forming industries

There is a clear and important distinction to be made between production lines which assemble components into final products and those which actually form components or final products. Of course, individual factories may include both activities: for example, automobile plants both form components (pressing bodies and casting engine blocks) and assemble these together with other bought-in components, into final products. But because there are basic differences in the nature of the technology used in forming and assembly, there are variations in the types of automation which are involved.

Assembly activities are essentially similar in all sectors, differing predominantly in relation to scale. They comprise of three separate stages – picking up the components, positioning them in the appropriate places and subsequently joining them together – and involve five major assembly techniques namely welding, brazing and soldiering, adhesive bonding, mechanical fasteners and stitching (Advisory Council for Applied Research and Development, 1979). More substantive differences however arise in relation to forming. Outside of dimensional products, where forming predominantly involves chemical reaction or extruding, there are five major types of different forming process, namely cutting, force-operations (including metal-forming machines), moulding, electro/chemical operations and treatments (including painting) (Amber and Amber, 1964).

ACTIVITIES IN THE MANUFACTURING SPHERE

Hitherto we have explored those sectoral factors which have an important influence on the activities involved in this sphere of production. It is now appropriate to set these differences in the context of the wider range of activities which are involved, and in so doing, to define the major, common activities in this sphere.

As we saw in Chapter 2, Bell (1972) distinguishes between three components of automation. The first is the central forming process which is involved; the second relates to the physical transfer of the raw materials, components and final product; and the third involves the control of these two activities. To derive the full set of activities in the manufacturing sphere, we add four items: setting up equipment, testing the product, storeage and distribution. In Figure 4.2 we illustrate the nature of the various activities which we see as being involved in the sphere of manufacture and distribution. Some activities – such as the setting-up of equipment, the storage of raw materials and final product, the inspection of final product and its distribution – occur in all sectors. Others – such as the core process itself and the method of handling raw materials, intermediates and final products – vary significantly between sectors depending upon whether the process is continuous or non-continuous, on the scale of operation and whether it involves assembly or component production. Each of these activities are of course, composed of subsets of more detailed activities. For example, Cook (1975) disaggregates the activities of a forming machine in the following way

> In the operation of a general-purpose machine there are a certain number of functions that must be performed either manually or automatically: (1) move the proper workpiece to the machine; (2) load the workpiece onto the machine and affix it rigidly and accurately; (3) select the proper tool and insert it into the machine; (4) establish and set machine operating speeds and other conditions; (5) control machine motion, enabling the tool to execute the desired functions; (6) sequence different tools, conditions and motions until all operations possible on that machine are complete; (7) unload the part from the machine (p26).

Fig. 4.2 Activities in the manufacturing sphere

Process ('dimensional products')		Non-process ('integral products')			
		Assembly		Component production	
Setting up		Setting up		Setting up	
raw material feed ↓	Continuous flow handling	component feed	Discontinuous handling	raw material feed	Discontinuous handling
processing ↓		assembling		forming	
inspection ↓		inspection		inspection	
storage ↓		storage		storage	
distribution		distribution		distribution	

TECHNOLOGIES FOR AUTOMATING MANUFACTURE

Given the wide definition of automation which we are using, there have been a very large number of technological developments which have allowed automation technologies to develop in this sphere. For example when we consider Amber and Amber's lower orders of automation (Chapter 2) and focus on mechanical energy, we see the transition from manual to water power in the eighteenth century, from water to steam in the nineteenth century and the development of electric power and the internal combustion engine in the twentieth century. Or if we were to focus on materials technology we might observe the earlier importance of iron and steel (which supplanted wood), followed in the nineteenth and twentieth centuries by the emergence of chemical technologies. Both these sets of technological developments in transformation and transfer predominated in the period before the second world war.

However, as we observed in Chapter 2, the key advances which are facilitating the diffusion of automation technologies in the last half of the twentieth century arise within the realm of control devices, which are being revolutionized by the introduction of informatics technologies. This involves the convergence of logic (ie control), data processing and communications within electronics technology. Of course, each of these elements of informatics technologies as well as other components of cybernetics technologies such as sensing and switching, existed in the pre-electronics era before the invention of the transistor in 1947. Thus already in the nineteenth century primitive control devices were introduced particularly in relation to machine tools. For example, by around 1850 the turrel lathe had the ability to memorize a number of operations, but sequencing was manual. At the end of the century a sequencing capability had been provided by using cams, relays and switches. And then in the early twentieth century the copy lathe was introduced, providing a fairly comprehensive degree of control, albeit circumscribed by the prior manual production of a model. Switching technology began to mature in the 1930s at around the time of the embryonic development of sensors. Nevertheless, despite these advances the first truly cybernetic technologies, that is, those involving some form of feedback control, were developed to control anti-aircraft guns in the second world war since manual guidance systems worked too slowly (Bednavik, 1965). These technologies all operated without electronic components which, as we saw in Chapter 2, were only really developed in the 1950s.

Soon after the second world war the first of the digital numerical control (NC) technologies were introduced, initially operated by non-electronic valve technologies. The operation of digital logic and counting systems – which are the heart of numerical control systems – are described in Chapter 2. It is through the development of this numerical control technology that modern-day automation is proceeding and it is consequently necessary to consider the origins and progress of this technology in greater detail. But before doing so we should note that numerical control technology originated primarily in relation to machine tools, with the exception of numerical control automatic testing equipment in the 1960s which was associated with the growth of the electronics industry itself. However, all of the numerical control systems – whether in machine tools, robots, testing equipment, process controllers, transfer lines or storage systems – operate on a similar logic. As a major UK-government sponsored study on industrial robots concluded:

> A strong indigenous control system technology has not been maintained [in the UK] and is essential to exploit the applications potential, whether robot manufacture is started or not. *The manufacture of robot control systems is closely allied to the manufacture of control systems for numerically controlled machine tools* (The Siemens System M, for instance is used for both) [emphasis added] (Ingersoll, 1980, section 4, p. 2).

As we shall see, it is precisely because of this compatability between different numerical control logic and counting systems in different activities and in different spheres, that automation technologies are likely to prosper in the coming decades.

The history of numerical control technology (Lund, 1977; Noble, 1979) largely begins with the US Department of Defence in the 1950s, although similar developments also occurred at around the same time in the UK. The impetus for the development of numerical control came in 1948 from a small Michigan contract manufacturer of helicopter blades which found difficulty in the necessary machining of complex contours. Using a desk calculator, the technical director was able to determine work positions for the jig borer, and, although machining remained manual, accuracy was greatly improved. Emboldened by this the firm took on the production of wings for Lockheed aircraft, bringing in the Massachusetts Institute of Technology (MIT) as a subcontractor with the help of a grant from the US Air Force. After a while it became obvious that the task was more complex than had been anticipated and the firm withdrew leaving MIT in the US (and Ferranti and EMI in the UK) as the sole developers.

In 1952 MIT exhibited their first working model of a numerical control machine tool. This was continually improved so that by 1960, point-to-point (as opposed to the more complex continuous path) machines were available for a wide range of uses, often including automatic tool-changing devices. 1959 saw the first introduction of electronics in the control system and these rapidly came to supplant the older valve-based systems.

Critical to the development of automatic, programmable machine tools was the type of 'knowledge' which they embodied to make them work. In essence two alternative paths were available. The first, as incorporated in 'record playback technology' was based upon the machine being 'hooked' on to a skilled worker who performed the basic tasks, thereby teaching the machine. The second path, much more complex and abstract, involved the development of a software capability which allowed the machine tasks to be defined in an 'abstract' way, without directly copying the skilled worker. Each individual user then applied this abstract machining language to fit its own particular requirements. It is this latter path which became known as numerical control.

Why numerical control triumphed over record playback is a contentious issue and is discussed further in Chapter 8. The facts are, that largely through the backing of the US Department of Defence (which invested $62 m between 1949 and 1959 in funding the emerging numerical control technology, plus a further $30–40 m in subsidizing the first users), the numerical control route was chosen. Crucial to this was the emergence of a unified software language, since by 1960 over 40 different variants were in use or under development. The variant which came to dominate – only after heavy pressure by the US Department of Defence which specified which language it was prepared to allow customers to utilize – was APT (Automatically Programmed Tools) developed at MIT between 1965–69, and subsequently upgraded in the 1960s by the Illinois Institute of Technology Research Institute. APT operated by breaking down design data into a mathematical description along five axis via hundreds of thousands of separate instructions. Each user then filled in the 'flesh' of a particular design around the skeleton provided by APT.

What numerical control did for automation was to reduce the actions of machines to a common currency. True, various versions of this software came into existence (which were mated together via electronic devices called post-processors). But the existence of an essentially common system, built on the basic binary logic which allowed for the use of electronic control and communication systems, laid the basis for subsequent intra-sphere and inter-sphere automation. It was thus an absolutely key development in the history of automation technology.

This binary logic system, as represented by numerical control systems, is now being used increasingly in each of the activities in the manufacturing sphere (which were illustrated in Figure 4.2). In the control devices required to set up the operating sequence of individual machines or the production line as a whole, electronic systems are displacing electro-mechanical predecessors. For example, in the forming of glass bottles from molten glass, the process used to be controlled by a bulky, well-bedded revolving cam with protruding 'fingers' which activated pneumatic timing devices. These are now being supplanted by electronic 'boxes' in which the control-logic to set up machine operations involves no moving cams or pneumatic activators, but rather solid-state circuitry and electrical relay and switching devices. A similar process has occurred in the inspection-activity where automatic-testing-equipment (ATE) is now widespread, especially in the electronics industry itself. Its further development in other industries is being constrained by the lagged development of suitable sensing technologies (although, as we saw, many of these hurdles are being overcome) and thus represents a major challenge to automation-technology suppliers.

Robots are finding increasing application in a variety of activities. The term derived from the Czech word *robota*, meaning serf, was first coined by a playwright in 1921. It is now taken to describe a variety of different types of serf-like functions. At its most basic level, it refers to mechanical manipulators which are mere extensions of human limbs. More complex are machines which possess a 'memory' (now almost entirely electronic-memories) and can perform a variety of sequential functions. In the more advanced from these machines possess reprogrammable memories, allowing for changes in their functions. But the frontiers of technological development involves the incorporation of sophisticated sensing devices which enables the robot to 'see' an item and to react in an 'intelligent' way to determine optimal solutions. Thus robots of the most primitive type are being incorporated in transfer lines to load, unload, and feed parts and pallets into or out of machines. In general, the Japanese include such machines in their definition of robots – they made up 78 per cent of their total robot population in 1976 – whereas American and European firms tend not to. More complex movable sequence robots are currently being incorporated in particular forming applications, especially in welding and painting and also in automobile production lines. They are also finding significant use in remote-controlled transfer devices which move parts and sub-assemblies around the factory floor (we will describe one such system being used by Fiat later in this chapter).

Numerical controls are now well established in machine tools used in the forming activity, and represent perhaps the most 'mature' diffusion of electronics in the manufacturing sphere. Similarly some elements in the storage activity – especially those involving the identification of particular parts – have for some time also utilized digital logic and electronic devices. But the retrieval function has not penetrated extensively and warehouses which incorporate automated retrieval and storage systems are a relatively recent – albeit rapidly growing – phenomenon.

All of the above represent examples of intra-activity automation technologies and were a characteristic of the pre-1975 period. Then, the introduction of the inherently flexible minicomputer and microprocessor in the 1970s, combined with the common binary logic of numerical controls in different intra-activity automation technologies, provided the opportunity to introduce the first attempts at intra-sphere automation. First came the machine centre, incorporating automatic tool changers. This was a single machine which could perform a number of tasks previously undertaken by different machines (for example, drilling and cutting). The next development was the machine cell which grouped a number of machines together with a dedicated and inflexible transfer line. The final major development, in embryonic form in the early 1980s, was the flexible manufacturing system (FMS) where flexible transfer lines directed work in progress automatically to machines which were available for transforming or assembling the components. Through this progression from individual intra-activity automation to intra-sphere automation there has been the increasing development of direct numerical control (DNC) in which individual numerically control machines are controlled by a central computer or hierarchy of computers. Clearly direct numerical control technology is essential for the development of intra-sphere flexible manufacturing systems, as is the emerging local area network (LAN) technology which facilitates intercommunication between different numerically controlled machines.

THE DIFFUSION OF AUTOMATION TECHNOLOGY

The diffusion of numerically controlled technology in the 1950s was facilitated by the subsidies given by the US Air Force for the purchase of 63 new numerical control machine tools and the refitting of 42 older machines with numerical controls. But progress was slow. The heavy data requirements of the APT system required the use of large computers which, in those days, were expensive and unreliable. Moreover, the early numerical control machine tools were introduced by inexperienced users who, for example, handled tapes with dirty hands, failed to provide electrical shielding for control systems and made insufficient allowance for the susceptibility of early systems to vibrations. By the early 1960s numerical control machine tools had become more reliable, but it was only in the mid-1960s that demand took off in the USA, and after the mid-1970s that relative demand increased in the UK (Figure 4.3). By this time, however, other countries had begun to use numerical control machine tools as well. In 1978, as can be seen from Table 4.2, the rate of use in Germany and Japan had already caught up with or exceeded that of the USA. In the lathe sector the utilization of numerical control was even more advanced and grew very rapidly over the 1970s (see Table 4.3).

The growth of other numerical control machines, aside from machine tools, has really been a phenomenon of the late 1970s. This time however, as in the case of robots (Ingersoll, 1980) the pioneering producing firms were Japanese, German, Swedish, Italian and French, as well as American; and the utilizing firms were in the early years (ie in the mid 1970s) predominantly Japanese and Swedish. Evidence for the growth of robot production in Japan is given in Figure 4.4; however, it should be noted that the Japanese definition of robots is considerably wider than other countries since it also uniquely includes fixed sequence, non-programmable systems. In Table 4.4 the types of robots are catalogued for the 1974–9 period.

Table 4.2 Machine tool production by type, 1978 (m. Swiss Francs). Shares of total production of the countries listed (%)

	Germany	*Italy*[1]	*UK*	*France*[2]	*Japan*	*USA*[3]
NC cutting	19.9	5.2	5.2	7.5	26.9	35.4
Other cutting	30.9	10.6	8.7	6.4	20.3	23.1
forming	32.6	10.1	5.8	6.0	19.8	25.6

Notes:

(1) NC Drilling machines included in drilling machines. Likewise other NC machines not separately identified, therefore the Italian share of NC machine tool production is understated.

(2) There is some overlap between NC boring, drilling and milling machines, so these figures are estimates only.

(3) NC milling machines includes NC grinding and polishing machines, included in other NC machines elsewhere.

Source: D. T. Jones, 1980

Fig. 4.3a NC machine tools as percentage of annual US machine tools sales (1959–1975) by value and number

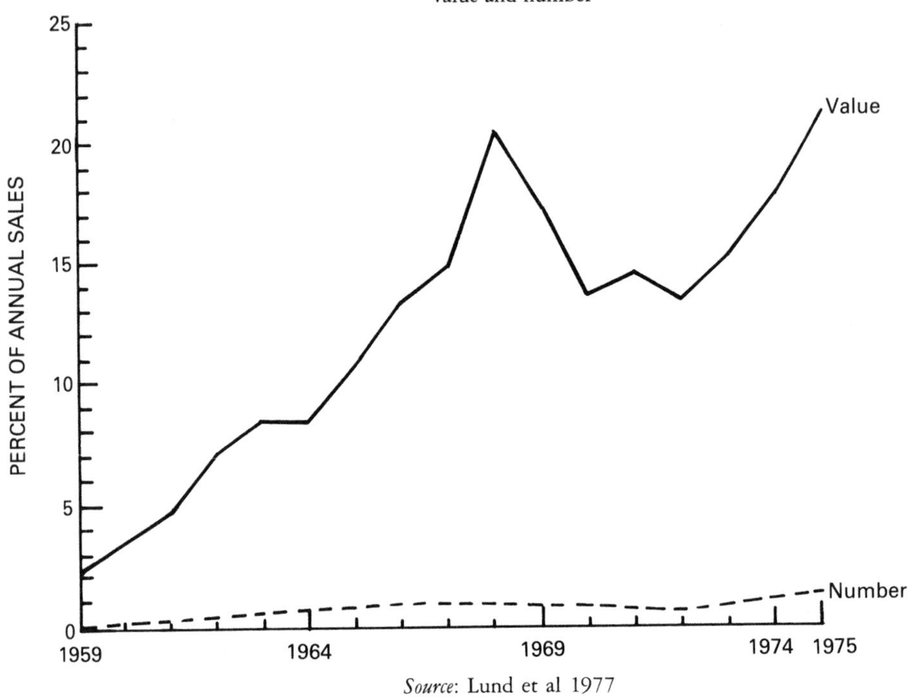

Source: Lund et al 1977

Fig. 4.3b Index of annual order intake by UK machine tool manufacturers for numerically controlled and non-numerically controlled machines tools (1971–1979)

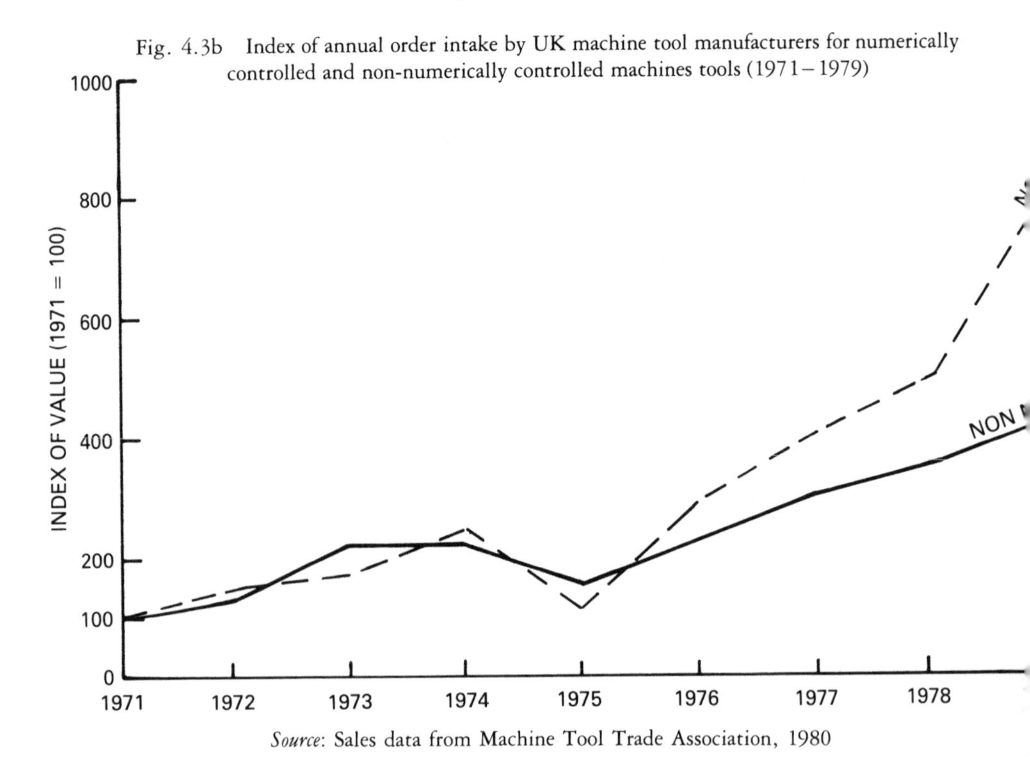

Source: Sales data from Machine Tool Trade Association, 1980

Table 4.3 Investment in NC lathes as percentage of investment in all lathes in major producing countries

Year	*United States*	*Japan*	*United Kingdom*	*France*	*Sweden*	*Germany, Fed. Rep. of*	*Italy*
1974	n.a.	n.a.	n.a.	n.a.	34.4	n.a.	n.a.
1975	n.a.	23.4	n.a.	n.a.	42.6	16.5	n.a.
1976	n.a.	28.2	18.6	26.4	41.6	n.a.	15.2
1977	n.a.	40.8	21.3	46.7	52.6	n.a.	n.a.
1978	n.a.	40.1	n.a.	n.a.	69.9	n.a.	n.a.
1979	n.a.	50.8	38.4	73.8	69.5	n.a.	n.a.
1980	56.5	n.a.	47.3	n.a.	n.a.	47.1	50.0

Source: UNCTAD 1982c

Fig. 4.4 Japanese production of robots. 1968–1979

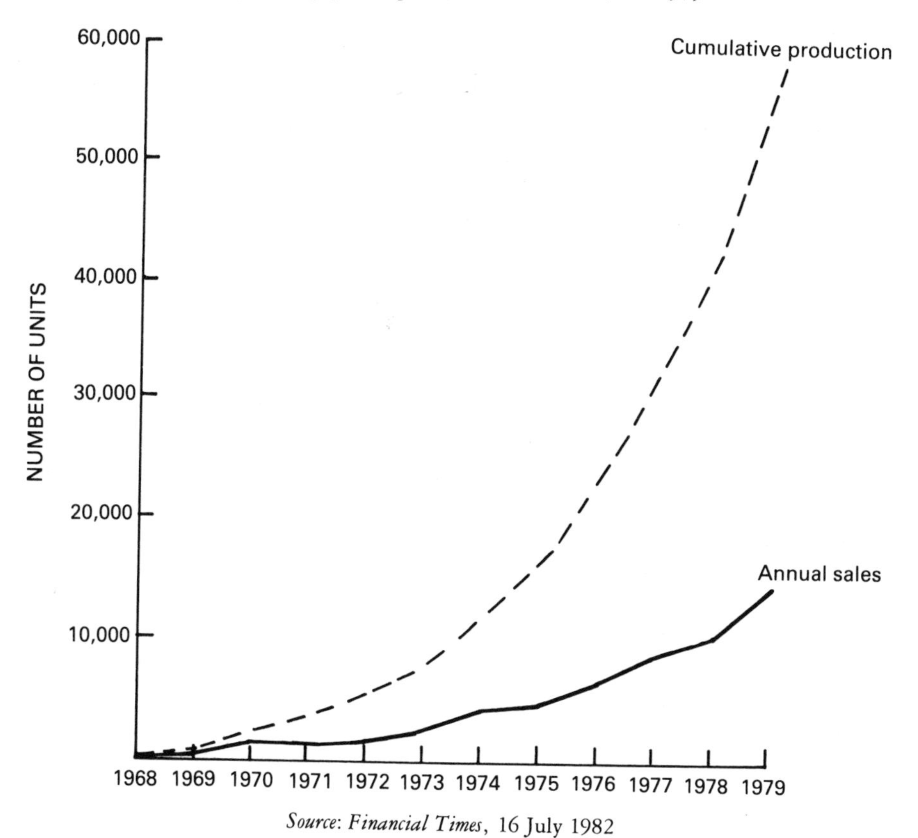

Source: *Financial Times*, 16 July 1982

Table 4.4 Annual production of robots in Japan, 1974–79.

Type of robots	*1974*	*1975*	*1976*	*1977*	*1978*	*1979*
A — Manual manipulator	713	772	697	1127	1,576	1,051
B — Fixed sequence robot	3287	3297	6199	6494	7066	10721
C — Variables sequence robot				425	652	1224
D — Playback robots	165	137	183	357	506	662
E — NC robot	1	0	6	11	25	89
F — Intelligent robot	1	12	80	199	255	788
Total	4167	4418	7165	8613	10100	14535

Source: *Financial Times*, 16 July 1982, taken from JIRA.

BENEFITS PROVIDED TO USER FIRMS

We see thus an increasing trend in the diffusion of electronics-based automation technology in the manufacturing sphere of production. This occurred especially after the mid-1970s when the long period of development of numerical control had resulted in reliable, durable and relatively cheap equipment. Clearly, though, the benefits this equipment provided to user firms must have been substantial, or otherwise the pace of diffusion would not have quickened so remarkably. We can group these benefits into four major categories, namely product quality, the flexibility of production lines and savings in unit labour and unit capital costs.

Improvements in product quality

We have seen already that the initial development of numerical control technology in the immediate post-war period was dictated by the needs of the aircraft industry for high-quality machining of curved wing surfaces. Only automated technologies offered the potential for adequate and consistent levels of product quality. These requirements were extended in the 1960s and 1970s to the electronics industry where automation has become a necessity for maintaining quality standards in production. For example, in the 1979–81 period Japanese electronic firms were able to make significant inroads into the US semiconductor market because their products were much more reliable. This followed directly from their use of more automated manufacturing technology: whereas around 37 per cent of US chips were assembled abroad in low-wage countries using labour-intensive techniques, the equivalent figure for Japanese firms was only 3 per cent. The American firms have responded to this situation by almost without exception planning to introduce automated assembling plants, often relocated in developed countries (Rada, 1982). A similar process of transition from labour-intensive to automated techniques – dictated as much by the necessity of improving product quality to cope with increasingly competitive markets as the desire to save production costs – has occurred in a wide number of other industries, such as in colour television production (Sciberras, 1979). The automation of forming, assembly and painwork, is also a constant tendency within the automobile industry where manufacturers search for product competitiveness as much as cost advantages.

Flexibility

Within technology currently available, the economics of manufacturing are such that there exists a constant tension between 'efficiency' (that is cost minimization) and flexibility. Einzig (1957), for example, reviewing the pace of automation in the 1950s, observed that

> 'One of the major technico-commercial handicaps is that in most industries automatised machinery is only suitable for production in long series. Owing to the high cost of its installation and of its adjustment, it is not a commercial proposition to apply automatic equipment unless there is a possibility of mass producing the same product over a fairly long period . . . The risk of a change in public tastes, necessitating an adjustment of the machinery soon after its installation, discourages managements from introducing automation' (pp21–2).

Even earlier, in the 1920s, Henry Ford installing the new production line for the low-cost Model T Ford offered the car to his customers as being available in any colour as long as it was black.

However, we noted earlier (Table 4.1) that around 75 per cent of manufacturing output occurred in small and medium batch production, suggesting that the necessity for flexibility made it difficult for many manufacturers to capitalize on the low costs of production offered by 'hard [ie inflexible] automation'. At the same time the inflexibility of the earlier numerical control machine tools confined the use of these automation technologies to industries where cost was relatively unimportant when set against the greater accuracy which numerical control made possible. Its use was dictated more by product-quality considerations, than the desire to automate batch production.

Then in the early 1970s three technical developments in the electronics industry – the invention of the microprocessor, the development of minicomputers and the continued reduction in hardware costs – provided a route for the wider diffusion of numerical control systems in manufacture, due to the possibility of introducing flexibility into their operations. 'Hard-wired', inflexible numerical control tools have now almost entirely been supplemented by programmable 'computer numerical control' (CNC) and 'programmable logic controller' (PLC) systems, both in individual machines (intra-activity) and groups of machines (intra-sphere automation). However, as is common practice we refer to all of these systems in subsequent discussion by the generic title numerical control (NC).

These developments in machine flexibility have a particularly important

Fig. 4.5 The effect of flexible manufacturing systems on automation and product composition

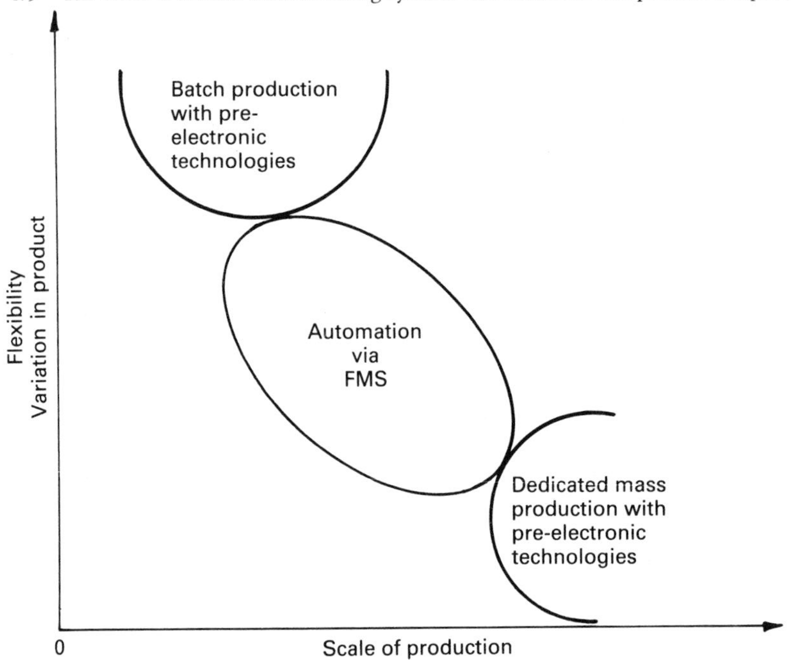

impact on the scale at which automation becomes feasible (see Figure 4.5). This is because the primary reason why dedicated hard automation transfer lines are inflexible is that it usually takes a great deal of time to change the settings on machines to produce products with different specifications leading to high 'downtime' of equipment. But when machines are controlled electronically the changeover times are often reduced very significantly. For example, in a new flexible manufacturing system being installed by one UK motor manufacturer, downtime during changeovers has been reduced from two days to 20 minutes, and at the same time a greater variety of engine types can be built. A second example can be drawn from the Boeing Aircraft Corporation which is introducing a new flexible manufacturing system to supplant its existing inflexible battery of numerical control machine tools which currently produces batches of sheet-metal prototypes with lead-times of 28–30 days. By 1984 Boeing anticipates a reduction of lead-time to 24–26 hours with economical one-off production of prototypes. Indeed the utilization of robots in the USA reflects the impact of flexibility on the batch size which automation makes feasible. As we can see from Table 4.5 in a survey of firms utilizing one-third of all industrial robots in America, there has been a significant move towards medium batch unit production as more flexible robots have been introduced.

Thus we see that a major benefit arising from the use of flexible automation technologies is that they facilitate the automation of batch production. A second and perhaps equally important benefit conferred by the technology, is

Table 4.5 The trend towards automation of batch and unit production: robot utilization in the USA[a].

Date of purchase	Batch size: Unit — No of firms	Batch size: Unit — No of robots	Batch size: Medium — No of firms	Batch size: Medium — No of robots	Mass — No of firms	Mass — No of robots	Total — No of firms	Total — No of robots
Pre-1976 users	1	10	4	28	7	112	12	149
Post 1976 users	7	63	2	5	6	15	15	83
Prospective users	6		8		4		8	

(a) Survey of 38 US members of Robot Institute of America, accounting for around one third of total US robot population

Source: Carnegie – Mellon (1981)

that it widens the variety of output in mass production lines. Take the automobile industry as an example. Unlike the early Ford plant which confined production to black cars, newly installed plants are now able to produce motor cars with individually specified variations (for example, in external colour, engine type and wheel-size) brought together automatically in the production line by electronically controlled intra-sphere automation technologies which match up the component feed with the assembly-equipment. Perhaps more significantly, the Japanese automobile industry has installed flexible production lines which are able to assemble different types of cars, as well as variations on a single theme. This will not only enhance the variability of output but also act to reduce economies of scale in production since a single plant will be able to produce more than a single type of vehicle. For example, the new Mazda plant can put together in succession a 323 front-wheel drive small hatchback, a 626 rear-drive medium saloon and an RX 7 rotary-engined sports car' (*Financial Times*, 19 October 1982). By contrast the new Ford Sierra and BL Metro automated production lines have concentrated on the labour-saving and quality-improving characteristics of the new technology. They are restricted to the assembly of a single type of car, albeit incorporating different variations.

It is important to note that these flexible automation technologies are not confined to robots. Consider, for example, the following flexible intra-sphere automated assembly line installed by Fiat, designed to combine automated transfer lines with assembly robots.

The LAM system at present uses 37 battery-powered trolleys travelling at up to 2.5 m.p.h. on magnetic tracks network totalling five miles. Each trolley, which collects a magnetic card carrying instructions for the precise engine to be built, can carry two engines. It delivers basic engine blocks and crankshafts to the first station, collects

them when the first phase of assembly is completed and moves then on the the next station.
There are a total of ten work 'islands', each of which has 12 work stations, of which one in each island is used for rectification. Each island is controlled by a microcomputer, which in turn receives its instructions from one of three computers which control the entire system . . .
Most important, however, output rates are no longer geared to the most complex unit in the family of engines being built . . . In the case of those [engines] going through the LAM system – there are 110 different detailed specifications . . .
At this level of complexity, Fiat insists, computer control of engine assembly is rapidly becoming not so much desireable as mandatory . . . (Griffiths, 1981).

Reductions in labour cost

It is in the reduction of labour costs that most attention is focused in the discussion of automation. Given that humans are (or, at any rate, are supposed to be) the central concern in social organization, this focus on the labour-displacing characteristics of automation in the manufacturing sphere is not unexpected. Yet, as we shall see, in Chapter 8, there are reasons to believe that the levels of labour-displacement are actually likely to be higher in other spheres. Nevertheless, there is abundant evidence that intra-activity and intra-sphere automation in this sphere are not insignificant. Consider, for example, the emerging inter-sphere automation lines such as those installed by Fujitsu Fanuc to produce industrial robots. Whilst the day shift consists of one hundred workers, the automated night shift involves only one overseer and one watchman. In the same firms motor-assembly, a FMS, working on a three-shift basis, produces 300 motors a day, whereas previously these motors were assembled manually on a single shift basis, with each worker producing 30 motors per shift. Fanuc calculate a 30 per cent reduction in unit costs (JEL, 22 September 1981).

It is possible to compute this job-displacement on a wider scale. Based on the responses of 16 major robot-using firms in the USA, accounting for over half of the country's total robot-population, Ayres and Miller (1981) estimated the effects of robots on labour-utilization in America's metalworking industries. They concluded that in assembly activities current non-sensor based technologies could displace around 10 per cent of the existing labour force, whilst the jobs of a further 30 per cent are threatened by emerging sensor-based robotization. The reduction in actual numbers employed is in itself an important benefit for user-firms (although clearly less beneficial for the workers displaced!) However, in addition, the automation of batch products offers a further benefit to these firms, in that it reduces the skill requirements of the labour force. Unlike assembly activities in mass production lines, which are not generally skill-intensive, batch production usually involves craft labour; their displacement offers proportionately greater 'savings' to their employers than that in mass production lines, since craft-labour is invariably paid at much higher rates.

Table 4.6 Comparision of labour costs and robot prices in Japan (Ym)[a]

	1970	*1975*	*1976*	*1977*	*1978*
Total labour cost per year	0.989	2.303	2.504	2.938	3.038
Mean price of playback robots	11.790[b]	11.120	11.01	11.9	11.1
Ratio of playback robot/annual labour costs	11.9	4.8	4.3	3.8	3.7

[a] For breakdown of sales of different types of robots in Japan see Table 4.4
[b] 1971 prices
Source: *Financial Times*, 16 July 1982

Whatever the potential benefits offered to user-firms, the utilization of automation technologies will depend upon their price relative to labour costs. With wages showing a tendency to rise and the cost of automation technologies declining over the 1970s, the balance has swung increasingly in favour of automation. Consider, for example, the relative costs in Japan (Table 4.6) in which the ratio between robots-acquisition costs and annual labour costs fell significantly over the decade. Thus even if considered only in relation to savings in unit labour costs, the introduction of robots is clearly offering substantial benefits to user-firms.

Savings in unit capital costs

When considering the progress of automation technologies over the past three centuries we are accustomed to viewing it as a process of machines being substituted for labour, a situation usually referred to as the capital-intensification of production. The problem with this concept is that it usually conflates two separate ratios: the first is that between unit capital and unit labour costs, and the second is the ratio of capital to unit costs of production. Given the changing capital/labour cost ratio reflected in Table 4.6 for robots, there is no doubt that the new automation technologies are increasingly capital intensive; however, they do not necessarily always involve a greater capital cost per unit of output.

There are two reasons for this. Firstly, numerical control systems, with few moving parts, are often substantially cheaper than their pre-electronic counterparts. Consider, for example, the control system used in the manufacture of glass bottles which we described earlier. The earlier electro-pneumatic control system required well-bedded rotating cams and pneumatic activating devices, whilst their successors involve bolt-on cabinets with electric relays. The crucial difference lies in the ratio between software and hardware costs, and here there are indications that the production of automation technologies involves some economies of scale to spread software costs over a large production run. However, the marginal costs involved in actually manufacturing the electronic control devices will in general be lower, and with long enough production runs of similar types of equipment, the fixed investment costs may actually fall over time.

A second area in which the new automation technologies involve reduced unit capital costs arises from the intensity with which capital equipment is actually used. In the American metalworking sector, one estimate is that productive cutting time in small and medium batch production is on average around 6–8 per cent of potentially available time and around 20 per cent in mass production industries (Ayres and Miller, 1981). New automation technologies offer significant advances over this, for three sets of reasons. First, because workers are shielded from hazardous tasks, production can be speeded up, often by a factor of 50 per cent (Ingersoll, 1980). Second, with the introduction of automatic loading systems (as used in the Fanuc plant described earlier), many plants are able to move to second and third shift utilization. And, third, working capital costs can be reduced substantially. Japanese automobile plants work on the 'last minute' system in which inventory is held for only half an hour's production; this, at least in part, depends upon the availability of electronic automation technologies in storage retrieval and communication. Moreover, the move to flexible manufacturing cells, incorporating

Table 4.7 Potential percentage increase in output due to recouping non-productive time as a consequence of intra-activity and intra-sphere automation

	Percentage machine downtime		
	With current technology	*With robots*	*With intra-sphere automation*
Mass production			
Load, inload, noncutting	20	18	15
Workstation allowances	20	12	4
Inadequate storage	10	10	5
Tool change	10	10	8
Equipment failure	10	10	10
Mid-volume production			
Setup and gauging	22	15.4	7.7
Load, unload, noncutting	12	7.2	4.2
Tool change	22	20.9	18.7
Equipment failure	7	7	7
Idle time	12	12	9
Low-volume production			
Setup, loading and gauging	55	41.3	27.5
Idle time	9	9	4.5
Cutting conditions	9	9	6.8
Likely increase in output			
Mass production		11	39
Mid-volume production		14	55
Low-volume production		11	39

Source: Drawn from tables 6.16, 6.17 and 6.18, Ayres and Miller (1983).

group technology (which we discuss in Chapter 6), will also substantially reduce the degree of work-in-progress in the average batch-producing plant.

Ayres and Miller estimate the output gains arising from the utilization of robots and intra-sphere automation in the manufacturing sphere for mass, mid-volume and low-volume industries. They calculate likely improvements in relation to various factors interrupting the full-time operation of metalworking plants based upon 'informed judgements [which] have been reviewed by several industry experts' (Ayres and Miller, 1983, p278). In Table 4.7 we summarize these results, distinguishing between various sets of intra-activity automation and full intra-sphere automation. Despite the inevitable error in these 'informal estimates' (coming incidentally from one of the premium engineering universities in America), the likely increases in output in high, mid-and low-volume manufacturing are striking, that is 11 per cent, 14 per cent and 16 per cent from using robots alone, and 39 per cent, 55 per cent and 52 per cent from full intra-sphere automation.

INTRA-ACTIVITY AND INTRA-SPHERE AUTOMATION IN MANUFACTURE: AN OVERVIEW

How do we relate this discussion of emerging automation technologies to our earlier suggestion that there exist three different types of automation, namely intra-activity, intra-sphere and inter-sphere? The last of these types is the subject-matter of Chapter 6 which will try to draw together the links between the various automation technologies outlined in Chapters 3, 4 and 5. Here we will briefly consider the first two types in relation to the sphere of manufacture.

Any historical study of industrial (or for that matter agricultural) technology will point to the growing advance of intra-activity automation over the past three hundred years. Usually these accounts dwell on the changing nature of physical energy from water to steam to electricity to internal combustion engines and finally to nuclear sources. Yet it would be a mistake to confine the analysis in this way since, when referring to the various activities in manufacture (see Figure 4.2) it is clear that in some cases other technical advances were of equal importance, such as changes in material technology and control devices in machine tools.

Prior to the late-nineteenth century, however, almost all these advances in automation were intra-activity advances. Then with the onset of mass markets, particularly in the USA, the era of mass production began and this provided a major impetus to the emergence of production lines. Once the 'line' organization of production was introduced, the opportunity arose for linking together automation in different activities. First, simple control devices were introduced to link the speed of feeding components on the line to the speed of the slowest link in the assembly chain. Then, after the second world war, feedback mechanisms were introduced to control processes both within particular activities and between activities. The introduction over the past thirty years of numerical control technologies in different activities – each working with the same binary logic – has further increased the potential for intra-sphere automation.

These electronic control devices have already passed through two major stages of development and are in the process of being introduced into a third stage. First, in the late 1960s, numerical control was used to substitute capital for labour and to machine complex toolpaths. Then in the last part of the 1970s (and especially in the automobile industry) the introduction of minicomputers and microprocessors led to the introduction of flexibility into the system, based on limited intra-sphere automation in flexible manufacturing systems. The next stage, now being developed, is the full linking of these numerical control systems throughout the manufacturing sphere involving all of the activities, from storage, through handling, machine setting, transformation/assembly, testing and storage. But the final stage – that is the linking of this intra-sphere automation to similar systems in the other spheres – is the ultimate 'prize' and a subject to which we turn our attention in Chapter 6. But before doing this we shall complete our analysis of automation technologies in each of the individual spheres and thus focus on the coordination sphere of production.

CHAPTER FIVE

Automation in the Coordination Sphere

The Communist Manifesto, first published in 1847, opens with the epic phrase, 'The history of all hitherto existing societies is the history of class struggles.' The control of this relationship between capital and labour is the primary function of management, in its role as the representative of the owning, capitalist class. And technology, as we shall see, has an important role to play in this struggle for control.

Those primary issues will be considered in detail in Chapters 7 and 10. In this chapter we shall turn our attention to the subsidiary functions of management in the coordination of diverse activities in the modern enterprise and in the setting and execution of strategic goals. Of central importance here is the generation, manipulation and coordination of various layers and types of information, both within the enterprise and between the enterprise and the external world. As we shall see, the new 'information technology' which is now enabling management to cope more efficiently with the handling of information not only represents the development of intra-activity automation but, by interconnection with other activities within and beyond the coordination sphere, also serves as an important link in the development of intra-sphere and inter-sphere automation. But whilst our ultimate concern in this chapter lies with the emergence of these automating information technologies, it is first desirable to begin with a brief account of the historical development of management as a specialized profession. This is because the coordinating requirements of management, and hence the activities it involves, only acquire significance in an historical context.

THE ORIGINS OF MANAGEMENT

It was in America that professionalized management training developed earliest and it is for this reason that we concentrate our attention on that country. There are a variety of reasons why American firms were in the vanguard.

America possessed the largest and fastest growing market in the mid-nineteenth century and then in the latter part of the century it saw the earliest introduction of mass production techniques which were closely allied to mass distribution outlets. Thus, the hundred years between 1840 and 1939 saw the emergence of managerial capitalism in America, first, and then subsequently in Europe and Japan. As Chandler (1977) observes,

> . . . the modern business enterprise took the place of market mechanisms in coordinating the activities of the economy and allocating its resources. In many sectors of the economy the visible hand of management replaced what Adam Smith referred to as the invisible hand of market forces . . . The rise of modern business enterprise in the United States, therefore, brought with it managerial capitalism (p1).

In the earliest (pre-1840) phase the American firm showed few signs of specialization. Small markets, 'primitive technology' and costly transportation provided little opportunity for the separation of the three spheres of production. Owners not only managed their enterprises, but were also responsible for process and product development. The evolution of cheap communication networks (notably railroads and telegraphs) first provided scope for specialization of the managerial function. In particular the task of coordinating extensive railroad networks (in the 1860s) led to the development of middle management with established operational procedures such as accounting systems. For the first time management came to be divorced from ownership (that is, outside of the agriculture sector which in 1815 already employed over 18,000 salaried managers (Chandler, 1977)). This middle management, as Chandler shows, was primarily responsible for the coordination of the day-to-day running of these enterprises. However, the transition to a new phase of intensive competition in the railroad network in the 1860s pointed to the need for a wider-ranging upper managerial level concerned with issues of strategic planning rather than mere coordination of on-going activities. This saw an important step in which senior management came to be professionalized and in time it too came to be separated from formal ownership.

In the 1890s the problems of day-to-day management of manufacturing enterprises (to which the managerial procedures developed on the railroads had spread) came to tax middle management, particularly in the sphere of production. This, together with the allied need to control the emerging conflict between capital and labour led to the development of 'scientific management' (of which we shall hear more of in Chapter 8) and professional societies such as the American Society of Mechanical Engineers which was formed in the 1890s. It also saw the emergence of 'junior management' roles such as supervisors and clerks.

By the 1890s, therefore, a hierarchy of three different levels of management had been developed in US business, a so-called 'functional' specialization. But the last twenty years of the century also saw the rise of intra-industry vertical concentration, and inter-industry horizontal concentration of ownership and control, as the national market came to dominate almost all sectors. The established functional organization of management proved inadequate to deal with these complexities and an additional layer of decentralization was introduced. This accompanied the development of multidivisional firm structures on which the functional specialization came to be subsumed under divisional hierarchies (generally grouping similar products together), all reporting to a central board.

The final stage in the development of modern managerial structures came as a consequence of the internationalization of production which began to occur after the turn of the century (Chandler, 1977). Characteristically because of their inherent riskiness, firms tended to hive off initial foreign operations as separate entities outside the established divisional and functional responsibilities. But as the foreign operations grew, so too did the tendency to subsume them in the existing, multidivisional, functional organization of the enterprise (Stopford and Wells, 1972). Associated with this growth in scale and complexity was the increasing separation between ownership and control, in which management came to be less dependent on the diffused owners of most large, international firms (Berle and Means, 1932). The interests they pursued were dictated by the wider logic of expanding in an international, oligopolistic economy and the strategic choices they made often represented the interest of management (eg growth) rather than owners, who might have preferred greater dividends with lower rates of growth (Williamson, 1963).

THE PURVIEW OF MANAGERIAL COORDINATION

Management, therefore, came to be organized in a characteristic structure of divisions and hierarchies. Upper management became responsible for broad strategic choices and decisions, and was situated in the Head Office; middle management became responsible for coordinating short-run activities and efficient operation and was housed in the 'offices' of divisional or regional headquarters; lower management became responsible for executing these decisions and for gathering data and was housed on the shop-floor and in the 'office'. Crucial to the operation of this hierarchically organized mode of control was the movement of information which flowed not only 'up' and 'down' the hierarchy, but also within layers of management. In order to discuss the automation of this information flow, it is first necessary to describe the areas in which managerial coordination is exercised. These can be grouped into four major sets, namely purchasing, production, sales and finance.

Purchasing

A wide variety of purchased items are used in manufacturing, although clearly sectoral variations are overwhelmingly significant in determining the type and number of purchased inputs. It is not only services, raw materials, partially made-up and fully made-up components which have to be obtained but also capital and labour. While the broad strategic decisions on the sourcing of all of these purchases (and particularly of capital and labour) will be made by management at the highest level, the administration of wages, stocks and capital expenditures will inevitably involve various levels of middle and lower management.

Production

Management has three functions with respect to production. First it has to plan and make strategic choices on the products to be introduced and the processes to be used. With the existing social division of labour, these decisions are inevitably made by upper management. Middle management then has to schedule the organization of production itself. The efficiency with which this scheduling occurs, particularly when this involves the assembly of made-up and bought-in components, has a very significant impact upon unit costs, product quality and meeting delivery targets. Lower management's function in this is to implement these schedules in an efficient manner.

Sales

The nature of the tasks involved in selling the output of an enterprise will clearly be dominated by the type of product involved. In the case of public utilities (eg electricity generation) purchasing (eg of coal) is often in the hands of a single firm and management is merely required to implement billing procedures. In other sectors, such as consumer goods, management may have an important role to play in actively marketing branded products. While in yet other types of sectors, particularly those involving 'software technology' such as computer-aided design, extensive after sales cooperation is required between the seller and the buyer and this will involve the establishment of relevant service networks.

Finance

The mobilization of finance is clearly a role which is exercised by management at its most senior level. In addition there are a wide range of financial controls which have to be exercised at all levels including product costing, the pricing of goods exchanged within the firm and the organization of weekly, monthly and annual budgeting and accounting documents. These will require managerial support at all levels.

ACTIVITIES IN THE COORDINATION SPHERE

In the simplest, unspecialized enterprises of the pre-industrial revolution period (see Chapter 2) the management function was not differentiated from production or design. Information control and communication was consequently not a problem as the complete process of production was concentrated within a single or limited number of individuals. But as the enterprise became larger and more diversified, and as the management function became more specialized and hierarchically structured, so the requirement of recording events and communicating these observations, and the resultant decisions, became more complex. We have now reached the stage in which each of the areas of managerial coordination outlined above can only be executed efficiently if information flows are controlled and coordinated effectively. This involves the production of information, its storage, its manipulation, its presentation and its transmission. And, as Caves (1980) points out, the larger the enterprise the greater the amount of information which has to be handled and controlled.

> Given the supervisor's [ie lower management's] span of control, enlargement of the number of primary operatives requires a predictable increase in the number of supervisory levels; more vertical levels mean greater 'control loss' as messages get garbled while passing up or down the hierarchy, and the top coordination level experiences increased problems with 'bounded rationality' – its ability to absorb and act promptly upon all relevant information (pp66–7).

In this process a two-way flow of information is involved. Raw data and analysed data passes up the managerial hierarchy, and this is counter-balanced by the downward flow of decisions which require execution or requests for new or amended information from lower levels in the hierarchy. This gives rise to the five major activities in the coordination sphere, namely the production of information, its storage, its manipulation, its presentation and its transmission. Given the multidivisional and functional specialization of management, each of these five sets of information activities is involved in all areas of managerial coordination.

The production of information

For upper management to comprehend the progress of production, design and marketing, it has to have a clear idea of what is actually happening and this inevitably involves the production of information. For example, telling-clerks may count the number of items made, wage-clerks will record the hours of work involved, 'chasers' will record the progress of manufactured items as they pass through successive stages of production, accounting clerks will record the level of sales, warehouse staff chart the progress of deliveries and secretaries type reports of these, and other, records. All of these 'facts' are essential to efficient management.

Information storage

Some of this information will be of only immediate relevance (for example, a memo to remind an executive, or a supervisor, of a meeting) and can thereafter be easily forgotten. But more often it is important that information be stored so that future operations can be assessed on the basis of historical performance. Some examples can be given in illustration. Accounting evaluation depends on the comparison of current years with past years; the performance of a machine can only be gauged in relation to its past operation or the performance of similar machines; sales forecasting depends upon evaluations of market potential in the light of past operations; purchasing at the lowest prices depends upon comparison with what the firm paid in previous periods and in relation to information gathered from alternative suppliers. Thus the 'memory' of information plays a crucial role in the efficient operations of the enterprise.

The 'manipulation' of information

On its own this stored information is useless. It needs to be 'manipulated' in various ways to enable management to extract key observations before it can make decisions. This 'manipulation' can involve various degrees of sophistication ranging from simple tabulations undertaken by clerks with minimal training to complex statistical analyses based upon a large input of 'number-crunching' effort. This complexity may not only require the consideration of a large quantity of information, but also the tying up of different bits of information collected from various parts of the enterprise. Thus in the calculation of costs of production for example, management will have to take into account data gathered from various activities in each of the three spheres of production: from both design and draughting in the design sphere, from the manifold activities in the manufacturing sphere and from each of the different managerial activities in the coordination sphere.

The presentation of information

The next stage in the control of information lies in the presentation of this recorded, stored and 'manipulated's data. Almost always this involves some form of visual presentation – for example drawings, reports, tables and graphs – but in isolated cases information may be presented in a verbal manner. Of course, in most cases these manifold forms of presentation are included in a single document, often accompanied by a verbal address to a group of colleagues.

Information transmission

Finally, in most cases it will be necessary to transmit this information (raw or 'manipulated') and the resultant decisions to other parties. It is significant that in the modern enterprise most of this information tends to be communicated within the firm. According to a study conducted by the British Post Office, over 50 per cent of information flows in the large, modern firm occur within the head office itself. A further 40 per cent arise between and within other divisions of the firm and only 10 per cent occur in relation to the external world (Walsh et al, 1980).

The relationship between these five activities in the coordination sphere and

Fig. 5.1 Activities in the coordination sphere

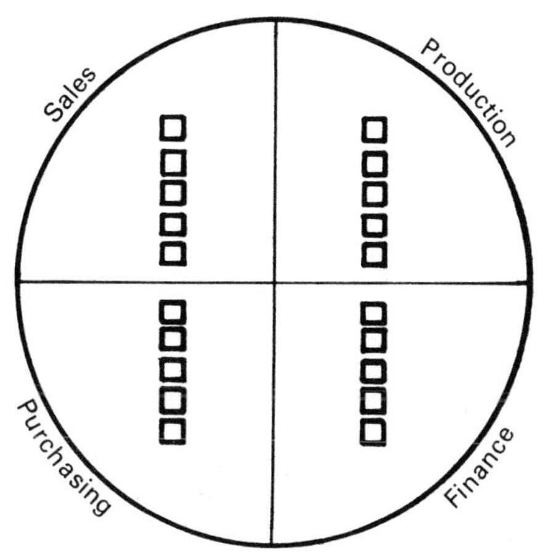

☐ Activities Information production
Information storage
Information manipulation
Information presentation
Information transmission

⊕ Functions Purchasing
Production
Finance
Sales

the four major functions of management is shown in Figure 5.1. In each of the functions, the various sets of activities are involved.

TECHNOLOGIES FOR AUTOMATING COORDINATION

Until very recently the technology available to execute each of these activities changed very little over the years. For example, according to one study (quoted in Rada, 1980), even in the most recent 1969–79 period, the growth in productivity of office staff in Europe (4 per cent over the whole period) lagged significantly behind that of industrial workers (80 per cent). Thus there is clearly considerable scope for the automation of information activities and the emergence of new electronic automation technologies over the last few years has lead some observers to predict radical changes in the organization of office (ie information) work. For example, in 1979 the UK National Computing Centre undertook a survey of leading decision makers in industry, government and academia to assess the impact of electronics on UK industry (NCC, 1979). They concluded that:

The office is where, to begin with, more changes are going to be made than anywhere else. At present the office employee is backed by less than one-tenth of the value of capital equipment behind his or her counterpart in manufacturing industry or agriculture, and this will change rapidly (p33).

Before we consider the emergence of these automation technologies it is worth reflecting on the nature of traditional office technology. Firstly, as we saw in the earlier section, it was only in the latter part of the nineteenth century that American management began to professionalize the gathering of data. The typewriter was first produced commercially in 1873 when its inventor, after failing to find sponsors, successfully approached an American arms producer (Remington) which had failed in its attempts to diversify into the production of agricultural machinery and sewing machines. But after the introduction of the typewriter, and then subsequently the telephone as a means of communication, almost no further automation occurred in any of the activities in information control, except insofar as this involved organizational changes, such as the introduction of typing pools. They remained overwhelmingly paper-based and labour-intensive: information was recorded manually, stored in bulky files, processed manually by various grades of clerical and professional skills, presented in manually typed reports and largely communicated by memos and letters.

Two major sets of automation technologies began to diffuse in the 1960s and the early 1970s. The first of these were mainframe computers which were initially used in information-manipulation activities such as payroll calculations and stock control (in fact the first commercial computer was introduced by a British bakery firm, J. W. Lyons, in 1952). These spread rapidly in large firms and led to the 'automation scare' of the 1950s and 1960s in which various commentators predicted widespread job-loss. However, this early phase of computer-use saw few cases of widescale labour-displacement, more often it led to an improvement in data availability and the generation of needless information (the GIGO – garbage in, garbage out – phenomenon).

At around the same time dictation equipment began to diffuse widely (over 600,000 machines were sold in the US in 1975 alone) and this allowed management to change office organization and increase productivity by introducing typing pools in which 'faceless' secretaries (almost always female) typed up information gathered by (male) management.

The wider diffusion of computers was hindered by four factors. The first was the inexperience of users who took time to master the new methods of organization which this automation involved. Second was the cost and unreliability of the computer hardware, and third, software programming skills were generally in short supply. Finally these early mainframe computers operated in a batch mode, that is, information was fed in, together with software programs, and only after a delay (often counted in days or weeks) did the computer produce its output.

All this changed over the 1970s. Most significantly the introduction of minicomputers and then microprocessors began to have an impact by lowering costs, increasing reliability and allowing for interactive (ie immediate response) rather than batch use. Electronics technologies began to diffuse independently in each of the various information activities and often in similar activities within each of the four different areas of managerial activity. For example, minicomputers were introduced separately to analyse information in purchasing, in production control, in sales and in financial controls. Consequently the era of distributed data processing succeeded the earlier phase of centralization, and provided the potential for intra-sphere automation (the so-called 'electronic office'). But before we proceed to examine this wider phase of automation, it is first instructive to chart the nature and progress of some of the intra-activity automation technologies which have diffused, or will diffuse, in the control sphere of production.

Information gathering

Traditionally information was fed into storage systems via manual recording on paper. Then came the computer which in the early stages required the transcription of this data on to coding sheets which were subsequently punched on to computer cards. (The labour intensity of this information-gathering system partly explains why computers did not *initially* have the employment-displacing effect predicted by some observers.) At around the same time the diffusion of dictation equipment began to displace some of the manual drafting of the earlier system. More recently, however, we have begun to see the emergence of embryonic forms of intra-sphere and inter-sphere automation in which information is directly communicated by machines without any human intervention, for example, in feedback-control systems in power plants, furnaces and other equipment; in earlier phases, humans would have monitored machines, and recorded their performance prior to taking action. And, as computer-based systems have become interactive, there has been a corresponding improvement in the ability to put information into the system without any intervening coding and punching.

This is now the state of the art of automation technology in this activity. The manually transcribed nature of earlier information-gathering systems has come to be replaced by the entering of information directly into storage networks by humans using computer-terminals or directly by machines. Where the technology appears to be going in the future is to enter data by voice. Already one firm, Calma, offers a computer-aided design system in which two hundred basic instructions (for example, 'draw straight line') can be offered verbally. An American robot producer, Unimation, offers a system in which the pace of operation can be verbally controlled by the operator, and various Japanese office-equipment manufacturers are predicting that by 1990 many of the information-entering activities will occur by means of voice-input (*Business Week*, 14 Dec. 1981). Indeed this is one of the primary objectives in their development of the fifth generation computer.

A second major development in information-entering technologies now occurring is the introduction of digital (see Chapter 2) recording techniques in which the recording equipment processes the sound in the same binary form used by electronics systems. This, as we shall see, is laying an important foundation stone for the coming phases of intra-sphere and inter-sphere automation.

Information storage

Information has traditionally been stored in a paper-based format. Then in the 1960s and 1970s a new technology allowing for photoreduction and reproduction became available and made it feasible to store much information in a reduced, photographic mode as represented by microfiche technology. At the same time the cost of storing information in electronic hardware fell dramatically, declining at around 30 per cent per annum since 1959 and shows every sign of continuing to decline at the same pace over the coming decade (see Chapter 2). This, together with the associated ability of computer-based systems to gain access to electronic storage files quickly, is beginning to make feasible the idea of electronic files in which the storage of information changes to a digital format (called computer output microfilm, COM, in the jargon) stored in paperless electronic systems. There are a variety of different storage technologies, some of which are built in to computer hardware (so-called 'core memory') and others which are available in a back-up form. While the actual system used in each case reflects changing technologies, changing prices and the type of software (so-called database management systems, DBMS) involved, there is little doubt that automated, non-paper-based information storage systems are becoming increasingly competitive with their traditional counterparts.

Information manipulation

The major new non-paper-based technology in this activity is the computer which, as we have seen, has evolved from expensive and cumbersome mainframe systems to cheap, small microprocessors. The growing power of these latter systems (either in 'computers' or 'calculators') has led to a reduction in the gap between these two processing technologies, which initially served separate markets. Indeed, microcomputers are currently one of the major areas of growth in the electronics industry and the very recent development of 'universal machines' which can 'read' other computers built with similar hardware, has begun to push the price of business computers below $1,000. In the face of the overwhelming dominance of electronic technologies in automating the analysis of information, it should not be forgotten that there was an intermediate phase of automation in which the efforts of humans were supplemented by mechanical calculators and slide rules (and in Japan by the even earlier use of the abacus). These mechanical calculators, in which IBM and National Cash Register, now two of the largest computer firms, first grew to dominance, were almost entirely eclipsed by the calculator 'revolution' of the 1970s which followed the development of the first microprocessor in 1971 (see Chapter 2).

Presentation of information

One of the major automation technologies in this activity has, until recently, not been one which uses electronics (and digital logic) as its central component. It has involved photoreproduction techniques which have come to be called, colloquially after the firm which pioneered its development, xeroxing. This technology spread widely in the 1970s and in most developed economies has displaced the earlier phases of mechanization (carbon-paper and stencilling duplicators). By 1979 over four in every ten US manufacturing establishments (of all sizes), possessed their own copiers.

More recently, Japanese firms have begun to erode the dominance of the established American suppliers by introducing cheap, small machines. And at the 'Upper' end of the market, new electrostatic and laser ink-jet systems have been introduced to speed-up and improve the quality of the reproduction processes. Associated with these changes in printing technology, a significant new development has been the introduction of digital, 'intelligent' copiers. These are based more extensively upon electronic components, rather than photoreproduction, and will have the ability (as we shall see in later discussion) not only to 'remember' the copied objects, but also to directly translate the digital signals of other electronic systems (for example, word processors and televisions) into printed output.

A second major technological development in this activity of information processing, has been the television screen. Initially developed for the military and consumer markets, the visual display unit (VDU in the jargon) has begun to penetrate the office. Currently it is most often linked with the entering of information into, and interaction with, information-processing computers, but it is now increasingly associated with the introduction of word processors. These word processors, are likely to be one of the most significant new office technologies in the near future. As such it is worth recapping their history a little. As we saw, typewriters were first introduced in the late nineteenth century, but they remained manually driven until the 1950s when electric motors were introduced to speed up the process and to lighten the physical load on the typist. In 1964 IBM broke new ground with its 'selectric' typewriter which provided an external electronic memory and which allowed the typist to edit the script. This early start was cumbersome, but provided the first link between typing and digital, electronic technology. In the 1970s the microprocessor provided the ability to reduce costs and enhance the editing capabilities of these 'intellectual typewriters' and the market began to expand rapidly after the mid-1970s. An associated development has arisen out of the capability of electronic control devices to outperform electromechanical ones and the resulting substitution of electronic typewriters for electric ones. Not only do these electronic typewriters have fewer working parts (six or seven modular parts versus 2,500) and are consequently becoming cheaper than their electric counterparts, but also they can easily be provided with electronic memories and thereby perform similar functions to custom-built word processors, which are in effect electronic typewriters with television screens. Those manufacturers such as Remington and Adler (now a subsidiary of Volkswagen) who have been unable to introduce this new technology early enough face extinction.

Information-transmission

The introduction of new information-transmitting technologies is providing an important building block for the automated electronic office. This is particularly evident in relation to communications with the external world and follows from technological developments in both the communications (for example, satellites and fibre optics) and electronics sectors. This allows not only for faster and cheaper communication but also for the introduction of new facilities. For example, the large 'band width' of signals required to produce a television picture makes it impossible to send these, and other, signals down conventional copper telephone lines: but the introduction of fibre-optic lines removes this restriction and provides scope for multi-form communication in the same network (for example, televisions, telephones and computers).

Finally, in assessing the overall diffusion of automation equipment in the coordination sphere of production, it is often argued that the Japanese language poses problems which makes it relatively difficult to introduce office automation technology. Nevertheless, whatever the relative rate of diffusion of this equipment in Japan, the absolute rates (as represented in Figure 5.2) represent a significant penetration of automation technologies into this sphere of production.

Fig. 5.2 Percentage of Japanese enterprises using automation technology in the coordination sphere

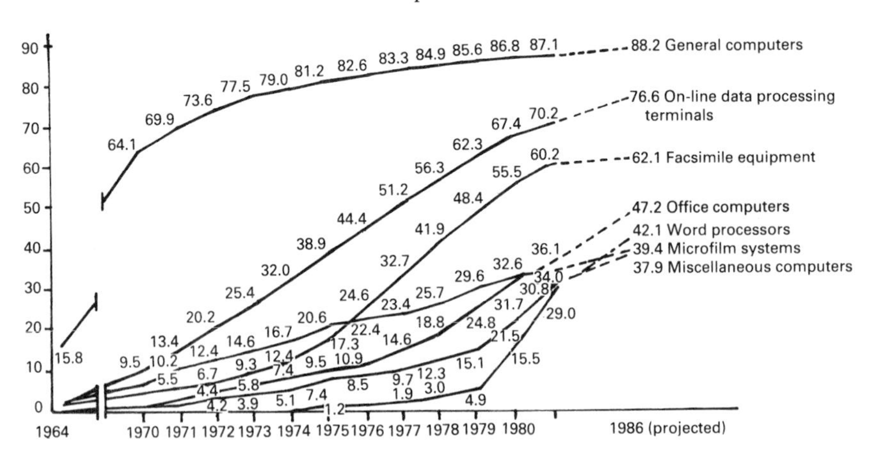

Note: The figures in this chart represent the proportions of firms among 100 enterprises surveyed installing respective types of equipment in the years indicated.

Source: Japan, Foreign Press Center 1982

THE BENEFITS PROVIDED BY OFFICE TECHNOLOGY

Intra-activity automation

As we noted above, most independent observers and equipment producers believe that the introduction of automation equipment in the coordination sphere of production will be a major area of growth in the near future. This belief is based on a combination of the low productivity of most information workers and the emergence of cost-effective automation technologies in recent years. So far, however, the technologies which have been introduced into information-related activities have been largely isolated in character, that is, they have been limited to intra-activity automation.

A number of studies are beginning to emerge on the benefits provided to the innovating enterprise as a consequence of using these automation technologies in this limited manner. These benefits can be grouped into three major areas, namely more efficient control, better quality output produced more rapidly, and reductions in cost. We briefly consider some of this evidence.

The *information control* requirements of modern enterprises, some of which employ over a quarter of a million people, are often overwhelming. One study estimated that on average in the US in 1976, 5000 copies were made for each of 15 million secretaries, stenographers, typists and clerical workers (Burns, 1980). Control of this information explosion is consequently a major task and the larger the enterprise the greater what Caves calls the 'control loss'. One way in which firms have coped with this problem has been to enforce decentralization which, despite its overall advantage in enhancing flexible responses, often only exacerbates these problems.

The introduction of computers has begun to stem this 'control loss'. In a study of 19 of the largest American life insurance groups (which are heavy 'information users'), Whisler (1970) concluded:

> With a small number of exceptions, companies reported that the effect of using computers is to push decisions to a higher level of organisation (p76).
> The perpetual problem of information overload that existed in precomputer days provided a compelling reason for breaking up the overall task of assessment, evaluation and control within a company into reasonably manageable subsystems (p98). Given the typical pyramidal structure of business organisations, this intergration [following the introduction of computers] results in shifting control up higher in the organisation than where it was formerly located (p99).

These observations on the centralizing effect of introducing computers were made after studies conducted in the 1960s, since when the capability of computers has increased rapidly further underlining the benefits to be obtained. In the future we can anticipate the marriage of computers with 'artificial intelligence' theory, a project which the Japanese computer industry and government are attempting to push through with the development of the fifth generation computer for the 1990s. Such computers are supposed to have the ability to evaluate raw data and to provide information selectively to users, rather than to swamp them with data. If achieved this will act to further the centralization of control in the hands of senior management.

It is difficult to chart changes in the *quality* of presentation in any systematic way. But the editing capability of systems possessing 'memories', allied to the introduction of visual (for example, business graphics systems as well as photocopying) and audio-visual technologies has in general vastly improved the quality with which information can be presented. A glance at the output of reports typed with traditional typewriters and word processors illustrates this well. The latter, for example, offers the ability to square-off all lines whereas with the former, the end margin is jagged and untidy. Moreover, with word processors, the author can play around with alternative font styles and spacing, to experiment until satisfactory copies are obtained. These quality enhancements are not merely graphical: the ease of editing makes it possible for authors to refine analyses at the last minute, without affecting either the cost of reproduction or the speed with which the document is made available. At the same time, particularly as the minicomputer and microprocessor have furthered the progress of interactive use, the new automation technologies have also increased the speed with which information can be prepared and presented.

The improvement in *cost-efficiency* is difficult to chart in any consistent way since there are so many different ways in which the new information technology can be used. For example, the evidence on word processors is mixed. Some observers argue that it is only cost-justified in particular circumstances – in one study when over 50 per cent of manuscript lines were amended, there were no productivity improvements, but when 23 per cent of lines were revised the productivity gain was over 100 per cent (Central Computer and Telecommunications Agency, 1980). Other observers (for example, Rada, 1980) are less equivocal with regard to cost-efficiency. Moreover, the rapid growth of sales of word processors (over 50 per cent pa) suggests that a large number of users are convinced of their financial feasibility. However, the continuing reduction in the price of these electronics-based systems suggests that even if automating information technology is not currently cost-efficient, it will become increasingly so in the future.

The evidence on new communication technologies is less equivocal. As a consequence of the superiority of new technologies the viability of traditional

Table 5.1 Cost of alternative transmission channels ($)

Type	*Cost ($)*
Internal telephone line	0.74 per 6 minutes
Facsimile*	1.97 " " "
Teletype	2.42 " " "
Internal memorandum	4.55 per page
US mail letter	6.41 per page

* Facsimiles are 'intelligent copiers' able to read, and photoreproduce, digital messages sent over the telephone line by other digital-logic systems.
Source: Adapted from Burns, 1977.

Table 5.2 Capacities of different transmission technologies

Technology	*Speed (millions of binary signals per second*	*Number of phone circuits*	*Cost per phone circuit ($1978)*
Conventional cable	5	500	200
Coaxial Cable	300	30,000	30
Terrestial-microwave	1,000,000,000	100,000	15
Satellite	1,000,000,000	100,000	30
Fibre optics	1,000,000,000,000	100,000,000	Unknown but very low

Source: Drawn from Cawkell (1978, reprinted in Forester, 1980).

information transmitting technologies is rapidly being undermined. Already in the mid-1970s the economies had begun to change in the USA as shown below in Table 5.1. The advent of satellite and fibre-optic technologies (Table 5.2) will further enhance this cost-competitiveness.

Intra-sphere automation

So far most of the automation technology introduced into the control sphere has largely been of a limited intra-activity nature, for example, word processors, photocopying machines and visual display units to enter information into computers. The gains to innovating firms appear to have been substantial, accounting for the rapid growth of this sector.

However, we are only just beginning to see signs of intra-sphere automation in which these separate intra-activity automation technologies are linked together. IBM and Xerox, the two dominant firms in office technology, announced in 1981 that they would be able to offer a local area networking system (commonly referred to as LANS) to link these separate elements together. In the words of one of the major electronics firms

> local area networks are basically a sophisticated 'pipeline' device enabling data (and voice, in some cases) to be transmitted between various kinds of electronic machinery (eg terminals, printers, intelligent copiers, facsimile, desk-top work structures) in a problem-free manner. Such networks may come to the market from a computing orientation, from an office machinery orientation . . . or from a telecommunications orientation.

When this networking capability is mature we will then begin to see the emergence of the 'integrated workstation'.

> Ultimately the product which will do most to change the operation of the office will be an 'integrated workstation' – the entry and exit point for information on a digital, voice, data, text and image network. In its fully developed form, this will replace the typewriter, the office copier, the telex machine, the computer terminal and the desk calculator, and will provide the key terminal facility for electronic mail . . . It will also have access to public data banks and information systems . . . (National Computing Centre, op cit, pp33–4).

The linking of these different activities in the control sphere will of course be gradual. Some examples can be given in illustration. We can soon anticipate the direct mating of a photocopying machine to a word processor; there are already small business-computers which combine financial analysis with word processing; facsimile technologies are marketed which link word processors and photocopying machines by the use of telephone lines so that a user in one location can photocopy a document which is reproduced almost instantaneously on another photocopy machine in a different continent, and so on (called fax machines).

In the case of transnational firms (TNCs) whose operations are spread over many countries, product groups and employees, intra-sphere automation of information coordination plays a crucial role in optimizing operations. Consider for example the conclusions of a report prepared by the United Nations.

> The reasons for which transnational corporations have established computer-communication systems are manifold. Most importantly, transnational corporations with integrated global operations are complex organisations because of the geographical distribution of markets and corporate assets, the diversity of products and services which are offered, the variety of economic, fiscal, legal and political environments that have to be taken into account, and the sheer volume of transactions. To control such an organisation, to manage it efficiently, and to monitor its performance, requires a large amount of information, especially for the parent corporations in which much decision-making power is typically rested. Furthermore, this information has often to be transmitted and processed without delay and must be stored and up-dated continually for future reference. Some if it (as in the case of various reservation systems, credit authorisation and inventory control) must remain accessible virtually on an on-line basis, while other data (like financial transactions) must be validated immediately.
> In line with the growing need for information for the management of transnational corporations, the areas of application of transborder data flows [ie intra-sphere automation across national boundaries] range across most corporate functions: production planning, standardisation and co-ordination; financial control; research and development; customer and purchasing questions; legal reporting and employee relations. In many of these areas the data of foreign affiliates can be processed by a computer system servicing the entire corporate network . . . Thus, transborder data flows give transnational corporations greater capacity to maintain the coherence of the corporate system as a whole and to allocate more efficiently and effectively corporate resources (UN, 1981, pp13–14).

There are still some technological obstacles to be overcome in the development of full intra-sphere automation in the coordination sphere of production. But more importantly, a major constraint on diffusion arises from the 'social' context in which the technology is being introduced. In its full form it will not only change the nature of office work but also inevitably lead to displacement of labour in particular offices and probably also in aggregate (Chapter 8). The automation of the control sphere of production is therefore problematical, a fact recognized by many firms in the supplying industry despite their public optimism.

CHAPTER SIX

Inter-Sphere Automation – The 'Factory of the Future'

In Chapter 1 of this book we observed the emergence of crisis in the world economy and pointed to the response of capital which, in recent years, has been to intensify the drive to automation. General Electric, under threat in almost all of its major product areas refers to these automation technologies as the 'factory of the future', while the Japanese activities have been coordinated by the Ministry of Trade and Industry (MITI) which has launched a programme to develop the 'unmanned factory'. In Chapter 2 we noted the underdeveloped nature of much of the (now dated) analyses of automation and offered a three way classification of the concept, based on the historical evolution of the modern enterprise into three spheres of production. Then, in Chapters 3, 4 and 5, we focused on the progress of intra-activity and intra-sphere automation in each of these three spheres, noting that the recent diffusion of electronics into these various activities has given a major impetus to the ease and imminence of the development of the final stage of automation, that is inter-sphere automation. In this chapter we survey the feasibility of this final, complex stage of automation. But before doing so it is necessary to briefly recap the major conclusions of the three preceding chapters on the development of intra-activity and intra-sphere automation in the three spheres of production.

THE PROGRESS OF INTRA-ACTIVITY AND INTRA-SPHERE AUTOMATION: A SUMMARY

In *design* it appears as if automation took a long time to arrive. As recently as the mid-1970s, intra-activity automation, particularly in engineering industries, was largely confined to the use of mainframe computers to assist with design calculations; intra-sphere automation, linking various activities of design, was almost unknown. Only in the electronics industry and the defence – aerospace sectors did we see the existence of embryonic intra-activity automation, and even in these sectors it was only around the mid-1970s that intra-sphere automation became viable (for example in using CAD systems to develop the 'masks' for integrated circuits). However, in the latter half of the 1970s both intra-activity and intra-sphere automation arrived with a 'bang'.

The supplying industry grew at a phenomenal pace and even in the recession of 1981–2 heady growth rates were maintained as user-firms rapidly appreciated the benefits which this automation technology provided. Thus by the end of the decade systems were already in place offering almost full intra-sphere automation, except for the initial activity of product specification which, by its nature, remained a 'labour-intensive' activity undertaken in consultation with senior management as part of the process of organizing corporate strategy.

In the *manufacturing* sphere there is a much longer history of both types of automation. Indeed one of the first recorded cases of intra-sphere automation goes back to an eighteenth-century grain-mill in America and even more primitive systems almost certainly predate this in other sectors and in other countries. The significant events in this sphere of manufacture, however, occurred in the 1960s and 1970s during which electronic control systems diffused to a wide number of individual activities, ranging from numerical control (NC) tool-cutting devices and transfer lines to robots and automated warehouses. Their significance lies in the ability of electronic systems to reduce lead-times, improve flexibility (for example, in changing machine settings), reduce costs (especially that of labour) and improve quality.

Most importantly, because of their use of a common binary logic, electronic systems offer the potential for mating together various activities. Now although, as we have seen, intra-sphere automation per se is not a new phenomenom in this sphere (eg moving assembly lines), the new electronics-based systems offer the same advantages to intra-sphere automation as they do to intra-activity automation. The critical significance of recent automation technologies incorporating electronics is that whereas intra-sphere automation used to be confined to large-scale mass production sectors (for example, assembly lines in vehicle manufacture) it is now becoming increasingly viable in small-batch production. This explains the significance of existing technological developments such as flexible manufacturing systems and more ambitious ideas as embodied, for example, in Hal-Technology (which will be discussed later in this chapter).

Finally in the *coordination* sphere it has long been noted that even intra-activity automation has been slow to emerge beyond typewriters, dictation machines and photocopiers. But in the late 1970s a variety of new intra-activity automation technologies were developed such as word processors and electronic typewriters, electronic files, small business-computers, intelligent copiers and fax machines. Supplying firms are now (ie in the early 1980s) largely concerned with improving applications software and marketing these single purpose machines, although some are actively beginning to put them together in systems embodying intra-sphere automation technologies. Thus, whilst the development of full intra-sphere automation – that is the integrated, multi-purpose workstation – is still some distance in the future, lesser forms of automation are beginning to emerge such as those systems which store and process information as well as offering word processing facilities. So the appearance of full intra-sphere automation technologies is a matter of time, and the speed of development depends as much on the receptivity of using firms as on the availability of suitable automation technologies.

IS FULL INTER-SPHERE AUTOMATION POSSIBLE?

The process we have recorded reflects a significant speeding-up in the development and diffusion of automation technologies as electronics diffused downstream over the 1960s and 1970s. And as the real cost of electronic equipment continues to decline in the 1980s, as software-development expands (with possibly even 'automated software writing' becoming feasible for certain routines) and as economic pressures inducing automation are maintained, so we can anticipate that the pace and degree of automation in the 1980s will continue to increase, and that diffusion will widen. It is important, however, to be wary of projecting into the future too glibly – as one observer pointed out, if the rate of expansion in the training of scientists and engineers in the USA (following the shock of Russia's launch of its first satellite in 1957) had continued unabated until the 1990s, there would have been two scientists for every man, woman and dog by the year 2000! (Jahoda, 1980). Clearly, therefore, there must be some limits to the advance of automation. Three stand out in importance

Technological impediments to full inter-sphere automation

If we return to the individual activities in each of the three spheres of production (Figures 2.3, 2.4 and 2.5) it is immediately clear that some, particularly in the coordination sphere, can never be fully automated since they depend quintessentially on human decisions. These include strategic decisions with respect to product development, procurement of inputs, and marketing strategies. Whilst in each of these activities the input of human labour can be augumented by automation technologies – for example, the Japanese are hoping to include in their development of the fifth generation computer, a capability for the computers to make certain decisions now made by humans – this human input can never ever be completely phased out. Nevertheless, there is still scope for the automation of a very wide number of activities now involving humans.

Full inter-sphere automation is not, however, only constrained by the absolute necessity to include humans in some key decision-making roles, but it is also held back by the need to develop suitable subsidiary technologies which allow the automation technologies to function efficiently such as sensing and activating mechanisms (see Chapter 2). Similar observations have been made in relation to earlier periods of technological development. Freeman (1974), for example, shows how the move from batch- to flow-production methods in the chemical industry necessitated the development of components such as pumps, compressors, filters, valves, pressure vessels and instruments. Bright

(1957), in his pioneering study of automatic machinery in the manufacturing sphere, believed that automation would require matching developments in materials, production processes, factory layout and design of the product. This reliance on other matching technologies he suggests was not unique to the mid-twentieth century.

> Historical research in almost any manufacturing field shows that productivity improves through a succession of parallel but random advances . . . Each advance often is limited by a shortcoming in another area (p20).

Nevertheless, despite the inevitable bottlenecks which will have to be surmounted in the development of adequate materials technology and components, none of the observers of automation suggest that any of these bottlenecks stand out as fundamental technological obstacles to the advance of full inter-sphere automation.

This would suggest that the re-emergence of the single sphere enterprise, characterisitc as we saw in Chapter 2 of pre-industrial revolution manufacturing activities, is a distinct possibility. Inevitably it will take time and will be subject to a variety of obstacles. Yet in principle, it remains a technologically feasible prospect save for the few strategic decision-making activities listed above and other specialized tasks such as design, sensing and repair. (Although even here automation possibilities are substantial: for example, many control systems are self-diagnosing, or incorporate redundant duplicating systems which are automatically brought into use when primary systems fail). But if the technological boundaries are relatively low, the diffusion of full inter-sphere automation may yet be limited by factors of an economic and social nature.

Economic impediments to full inter-sphere automation

Two linked factors are of relevance here. The first concerns the economic cost to supplying-firms and society at large of developing these automation technologies. These costs are partly 'human' in that they involve very substantial inputs of programming skills. For example, some of the computer-aided design equipment suppliers, although only covering a narrow range of the market, had by 1980 each already incorporated over 1,000 person years of software development (involving in excess of 7 million lines of programming) in their applications software.

But perhaps more significant is the second factor which reflects the ability of user firms to invest in radically new automation technologies. As we saw in Chapter 1, the emerging crisis of the pre-automated era was associated with the slower growth of markets and a decline in the rate of profit. So the financing of the installation of automation equipment will be problematic and it is in this context that the tendency of governments to finance the purchase of new information technology and the restructuring of established industries such as automobiles, steel, shipbuilding and basic chemicals should be seen.

A secondary phenomenom of importance here is the trade-off between automation and scale. As we saw, earlier rounds of intra-sphere automation were largely limited to large-scale mass production sectors. And whilst the flexibility and low cost of electronics make it increasingly feasible to reduce the scale threshold at which intra-sphere and inter-sphere automation becomes economically viable in each plant (but not necessarily in each firm – see the following chapter), there will always be a tendency for the limited advantages conferred to very small-scale production batches to be outweighed by the costs which are involved in the development of these automation technologies. At the extreme it is highly unlikely that it will ever be worth gearing up a fully inter-sphere automated enterprise to produce a single product. And, here, there will inevitably be sectoral variations in which the greater the quality required, the shorter the desired lead-time and the higher the value of the final product, the more feasible will be the full automation of small batch production.

Social impediments to full inter-sphere automation

In theory automation has a great deal to offer society since it provides the potential to liberate us from the necessity to work whilst at the same time providing us with a high material standard of living. But the extent to which humans can be liberated from the tyranny of specialized work reflects not just the degree to which technology is able to relieve us from scarcity, but also the way in which society is organized. So the pace of diffusion of automation technologies will necessarily reflect the distribution of the benefits which their use confers. It was precisely this fear that the benefits of intra-activity automation would be distributed so unevenly which led the Luddite workers to destroy the new automated spinning technology in the early nineteenth century.

Having made this observation – which often leads to the too simplistic conclusion that because capitalism inherently distributes benefits less evenly than socialism, it will inevitably experience slower diffusion of new automation technologies – it is important to recognize the complexity of the historical and political factors which determine the diffusion of automation technologies. Thus, in observing the tendency of the Japanese to install systems-based inter-sphere automation technology more rapidly than their European (and especially British) and American counterparts, it is too easy to be led into explaining this by 'cultural' factors. More persuasive are analyses which situate the discussion of social attitudes to technical change in a wider, historically based context pointing, for example to the 'historic compromise' between capital and labour which has emerged in post-war Germany and in Japan, but not in Britain or Italy. In addition, the rapid expansion of markets has allowed some Japanese firms in the past to offer life-time employment, clearly making labour more sympathetic to the introduction of these technologies.

The point of this brief discussion is not to offer a determinate description and analysis of the social factors which will impede the spread of full, inter-sphere automation – for this will inevitably change over time and vary between systems – but to point to the fact that the diffusion of automation technologies must be seen in the context of a struggle. This struggle, as we pointed out in Chapter 1, is currently recurring between different types of capital and at the same time also between capital and labour: the degree and pace of diffusion will reflect this broader struggle for dominance and it is this subject which will be the central concern of Chapters 7, 8 and 9.

THE PROGRESS OF INTER-SPHERE AUTOMATION

Thus intra-activity automation technologies – most significantly those incorporating electronic control systems – are now widespread in each of the three spheres of production, and intra-sphere technologies are rapidly emerging as these individual electronics-based automation technologies are linked together. But what can be said with respect to the emergence of the final, full stage of automation in which automation technologies in different spheres are being linked together?

Hitherto, most literature on the subject of automation has been largely concerned with either the minute detail of individual intra-activity automation technologies, or with the overall, global impact of automation on society. However, emerging out of the production engineering literature is a set of views which although starting from a very different perspective – that is the primary orientation has been technological – has generated rather similar conclusions to our own. A recent study by Halevi (1982) draws together the state of the art of literature in this field and points to the likely picture of future technological developments.

Halevi is concerned with Hal-technology, derived from the Hebrew word 'Hal' which means 'all-embracing' and so characterized as:

> . . . a new, all-embracing computer-oriented technology . . . that views the manufacturing process as a single unique system (pviii).

The essence of Hal-technology is the development of inter-sphere automation (although Halevi obviously does not use this phrase) and flows from the recognition of the limited gains offered by intra-activity and intra-sphere automation. Thus, in relation to intra-activity automation, Halevi observes that:

> Development in the use of computers as an end to manufacturing have proceeded in a modular, disjointed fashion. Systems designed and developed to solve a specific problem as expediently as possible were necessarily limited in scope (p393).

This points to the need for intra-sphere automation. Here:

> The evidence shows that the economic benefits to be gained from integration [ie intra-sphere automation] far exceed those benefits directly attributable to individual development efforts [ie intra-activity automation]. This is particularly true in discrete part-batch manufacturing based industries because of such factors as the need to maintain both a flexible fabrication base and highly efficient controlled operation (p393).

However, Halevi observes that this intra-sphere automation has occurred in an unsystematic way, that is:

Integration of these systems [ie intra-sphere automation] has been attempted in some cases, but only as an afterthought. The resulting proliferation of disjointed computer systems tends to magnify manufacturing problems (p393).

All this points Halevi (and, he argues, production engineering in general) to the concept of Hal-technology, which we have termed inter-sphere automation. Thus:

Hal treats the manufacturing process as one interactive problem starting from engineering design to product shipment. It considers the manufacturing process as a nucleus of satellites rather than a chain of activities (pp393–4).

It is worth exploring the historical evolution of inter-sphere automation – as seen through Halevi's production-engineering eye – in greater detail. Halevi points to four stages in the developing use of computers in production, the logical outcome of which will be the development of inter-sphere automation in which the various activities in different spheres will not just be linked together in an ad hoc manner but will be systematically incorporated as sub-elements in a comprehensively organized system.

The first of these stages was the *application approach* 'to solve a specific problem, or to supply specific information to a specific department or person' (p7). These technologies were isolated ('stand-alone' in the jargon), but their superiority over existing technologies led to the rapid spread of computers in industry. However, it soon became apparent to management that the 'data required for manufacturing or by top management are usually a combination of data from several stand-alone applications' (p8) so leading to the second phase of *data entry integration*. In this stage, the embryonic form of intra-sphere automation, information came to be stored in a centralized, computerized data base. Although this provided economies, it increasingly led to problems in the control of updating information (given varied users), the non-overlapping nature of data (not all users required the same data) and a consequent degradation in the quality and ease of access to information. Perhaps more importantly in these mainframe-computer based systems information tended to be application-oriented rather than systems-oriented and this placed obstacles on the development of intra-sphere automation. Consequently, as from the middle of the 1970s, intra-sphere automation was built around *distributed processing*, the third of these phases. These interactive, minicomputer-based systems (as we saw in the case of CAD in Chapter 3) allowed for low-level intra-sphere automation and spread rapidly. (One US Corporation cited by Halevi had 35 small computers in 1975, 102 in 1976 and 150 by mid 1977). The problem with these distributed systems was that they were too narrowly intra-sphere in approach to satisfy upper management's desires to spread wider forms of automation. Although they were able to eliminate duplication and inconsistencies and to make information available to other users (usually in the same sphere) they were too particularistic to allow for the wider ultimate benefits of inter-sphere automation.

All this pointed Halevi to the need for an approach which would make inter-sphere automation possible. He argues that the individual sets of this all-embracing technology are already in existence and the task now lies in putting them together. However, he does make the very important proviso that for this to succeed it is crucially important that inter-sphere automation should not proceed in a narrow framework in which individual activities in different spheres are linked together in a half-hearted, piecemeal fashion. It is vitally important to recognize that production is a unified system with its own respective subsets rather than a conglomeration of related, but separate systems.

SOME EXAMPLES OF INTER-SPHERE AUTOMATION

It is important to recognize that inter-sphere automation is not a phenomenon of the future, but is increasingly one of the present. To illustrate this it is useful to point to a few examples to show that the gains are already being realized: these are drawn from an analysis of firms using computer-aided design (CAD) technology (Kaplinsky, 1982b).

In Chapter 3 we examined the development and diffusion of automation technologies in the sphere of design, notably with respect to interactive computer-aided design technologies. We also considered some evidence on the gains realized from intra-activity and intra-sphere automation obtained from a variety of American and British firms which had made use of the new design automation technology. However, a number of these firms were beginning to widen their horizons and to reap the benefits of inter-sphere automation. Indeed the substantial number of these utilizing firms were at pains to point out that as far as they were concerned, the major benefits from use were going to be felt in downstream activities in the two other spheres of production. Assessing the state of the art of these downstream benefits poses substantial difficulties, partly because of the difficulties in estimating the extent of these links, partly because they are gradual and emerging, and partly because of the very substantial variation in the nature of the potential links, which depend upon the sector, the product and the type of factory organization. Therefore, it is only possible to illustrate the types of inter-sphere links between CAD and activities in other spheres as an example of the sorts of competitive advantages which can arise from the introduction of inter-sphere automation technologies. For obvious reasons related to the earlier discussion in Chapters 4 and 5, we can distinguish between benefits related to information coordination and benefits related to manufacture.

Information coordination

The extent of information required by most firms which utilize CAD are generally beyond the processing capabilities of minicomputers, and consequently in the case of most user-firms, parts listings, bills of materials and ordering were organized around mainframe computers, separated from the CAD systems. While it is quite possible for these different computers to intercommunicate directly, as a general rule there is some form of human interface between the two systems as well as some form of duplication of data generation and processing. However, the more powerful mainframe computers have no such difficulty and in fact this integration of data processing and analyses is the specific philosophy of IBM CAD-system marketing as embodied in their COPICS (communications-oriented-production-information-and-control-system) systems (Halevi, 1980). Moreover, as the turnkey CAD vendors move to the more powerful 32-bit computers (which they are all currently doing), their ability to integrate CAD-analysis software and batch-information processing will increase. Thus the unification of the CAD database with these in the coordination sphere is an increasingly common phenomenon.

Once this is done, the degree to which firms will be able to reap substantial economies will depend upon their existing structuring of information. For example, one of the users (which designs and builds process plant) has already made substantial inroads into parts listing via their mainframe computer: as a consequence of having a more specific knowledge of their needs for parts, on-site contingency costs have fallen from 15 per cent to 5 per cent since they were able to keep a much smaller supply of components, and construction was less frequently held up by a shortage of crucial components. In other cases, the pre-CAD organization of information had been suboptimal and the installation of CAD systems in these enterprises had forced the downstream systematization of inventories. This led to a more structured organization of warehousing and parts coding. More importantly, it led to a reduction in the number of different parts held, as the systematization illustrated unnecessary proliferation of part-types. In particular it had given an impetus to the development of the 'group technology concepts' which we will discuss later in this chapter.

Downstream information flows need not be confined within enterprises, of course. A particularly pointed example emerges from the British motor component industry where a British automobile firm is building a new car under licence from a Japanese firm, and to its design. One of the major UK component suppliers complained that the specifications supplied by the Japanese firm for a particular part were unintelligible: 'instead of sending us a drawing, all we got was a digital readout'!

Manufacture

There are very many potential links between design and manufacture and the specific benefits available depend upon the nature of the process and products involved. We illustrate some of these to give a flavour of some of the potential benefits which arise.

Machine setting: the control mechanisms of numerically controlled machine tools utilize the same basic digital information as the CAD systems (although for this information to be utilized it has to pass through a post-processor which, in the absence of direct numerical control (DNC), prepares the paper tape for the computer-numerical-control (CNC) machine tools). Consequently it is not a complicated task to link the two systems. In most cases, CAD users have drawn links between CAD designs and numerically controlled milling and cutting machines; but in one firm (which is a sign of a future trend) specifications for the automatic testing equipment were fed directly from the CAD system and in another active plans were being made to link assembly equipment to the CAD system. The benefit of these links are manifold, including the displacement of machine operators and a reduction in errors in machine setting.

Production planning: once a unified (and accurate) database has been established in the CAD system it is possible for this information to be assessed by multiple users. This not only spreads information more widely through the plant but also many firms found that it speeded up the release of final drawings (in one case from three weeks to less than eight hours) and allows for a single, corrected master-design to be utilized, rather than the previously haphazard proliferation of incomplete drawings.

Materials saving: a particularly important benefit of CAD is a saving in materials due to the optimization of design and nesting (the name given to the programs developed to cut shapes out of a sheet of material). In one firm optimized designs had reduced the number of parts in a machine by 50 per cent; in another, CAD had made it possible to reduce silver utilization by 50 per cent in a process in which silver comprised 30 per cent of direct production costs. Benefits from CAD nesting programs are widely felt: one sheet-metal user had reduced wastage from 40 per cent to 26 per cent in its first-generation nesting program; its annual savings in sheet metal equalled the total annual wage bill.

Prototypes: four users concluded that as a primary benefit, more accurate CAD drawings reduced the need for and costs of manufacturing prototypes. In one particularly graphic case, an electronic instrument had been built for an aircraft (at a cost of $100,000) which did not fit into the space available in the cockpit; this required a complete redesign which the firm argued would not have been required had a CAD system been utilized in the first place.

Extra-firm benefits: a major and rapidly expanding field for CAD systems is 'piping' software which is used for 'interference checks' (ie whether pipes obstruct each other or other sets of equipment in a process plant). Despite the fact that by 1981 none of the existing turnkey CAD suppliers yet appeared to have had a fully mature 3D software package for this, (although all had them under active development) the benefits in construction were already being felt. In the past, a significant but unquantified cost in the construction of process plant was incurred in the on-site rectification of design errors which resulted in interferences. The elimination of such design errors is already beginning to have an impact in reducing construction costs, despite the immaturity of much of the available software.

Another example of a trend towards inter-sphere automation can be drawn from the recent re-equipment of a locomotive-building plant by General Electric in Erie, Pennsylvania. Faced with the potential revitalization of the US railroad system and the recent success of Japanese and European firms in supplying metropolitan transport systems with new equipment, GE is investing $316m in a new automated plant.

> Starting at the beginning, the design output of the engineering department will be passed on to the manufacturing engineers in electronic form, rather than as drawings, and will then move through materials control, which will automatically schedule and order materials and keep track of stock and production flows.
>
> All this information will come together in the factory in the host computer, which will contain in its memory details about how, when and what to produce. This in turn will send instructions to the computer-controlled equipment, such as numerically controlled machines and robots, which will actually do the job. Quality controls, financial data, and customer service records will also be plugged into the same system (Lambert, 1983).

It is significant that GE expect such substantial savings – 25 per cent in direct labour, 20 per cent in material costs, 20 per cent reduction in inventories – that it expects the investment to be self-financing through savings in working capital alone. Already the introduction of a flexible manufacturing system (ie intra-sphere automation) to produce a family of motor frames and gearboxes typically completes job-lots in sixteen hours with four people working, compared to the earlier system which employed seventy people on twenty-nine different machines and took sixteen days. As a 'bonus', parts quality is higher and floor-space is cut by 25 per cent.

FACTORS AFFECTING THE SPEED AT WHICH INTER-SPHERE AUTOMATION DIFFUSES

In earlier sections of this chapter we have considered whether full inter-sphere automation was possible and on balance decided that this was so. We have also given some examples as a flavour of the sorts of benefits to be realized from inter-sphere automation. However, it is important to be aware of the difficulties which might arise in the development of inter-sphere automation over the coming decade. In addition to the social and economic obstacles to full automation discussed earlier in this chapter, two more specific factors are relevent. These are the difficulties in capitalizing on systems-gains given the tendency towards creeping incrementalism and the difficulty in harmonizing and matching information generated in disparate spheres.

Incrementalism: the hidden dangers

Since automation technologies are still evolving and are generally optimized on the basis of experience, their evolution will of necessity be incremental, although nevertheless occurring at a rapid pace. Additionally not only are the technologies evolving but so too are organizational structures within which they are implemented. There is a great danger in this evolution that the same fate will beset inter-sphere automation as that which befell, according to Halevi, its intra-sphere predecessor, leading to suboptimally organized subsets of technological linkages which are unable to capture the full extent of potential systems-gains. In the end, as Halevi points out, efficency considerations will force enterprises to take a holistic view of the production process, but the path to this destination may be tortuous unless this global perspective is recognized from the outset. This recognition appears to exist in Japan; it also has penetrated individual American firms such as General Electric, IBM and Westinghouse who, as we have seen, have recently come to grasp the potential magnitude of the new electronics-based automation technologies. Thus, rather than adapting an existing plant, General Electric preferred to construct an entirely new and highly automated factory for the manufacture of washing machines. Attention is therefore drawn to the importance of seeing production as a system. The capturing of these systems gains is a decision to be exercised at the highest levels of strategic management (see Chapter 5). This involves two separate phases. The first is the *recognition* of the potential benefits to be derived from inter-sphere automation and of the dangers in trying to achieve these benefits via piecemeal adaptations to existing production systems. Halevi describes the necessary involvement of management on the following terms:

> for this stage to be successful, it must be understood that system integration is not a data processing slogan or technique. It is, above all, a management technique in which it is realized that the manufacturing process is neither a set of independent systems nor several sets of integrated systems, but one logical, overall system . . . without management involvement, the chances are next to zero (p12).

The second phase is the *power* which management has, as the representative of capital, to enforce the required changes in production technology and methods. For as we have seen (and will discuss further in Chapter 8) inter-sphere automation requires very different work practices and almost always involves much lower levels of labour. In most circumstances, particularly in the UK, Italy and France, these changes will be resisted by the workforce, many of whom will be affected in an adverse way. Thus it is an open question whether, even if management recognizes the need for changes in the production system, it will be able to carry through the necessary measures in an adequate way, or in the approriate time scale required to cope with the advances made by competitors.

Harmonizing information

As we have seen the history of electronics in production has been characterized by the gradual introduction of a variety of individual intra-activity automation technologies, each with its own data needs, data entry systems and databases. Whatever the capabilities of the physical hardware to link these various databases together (through, for example, the introduction of local area network systems (LANS) which mate together the digital output from a variety of separate electronics systems), unless the information exists in a format which allows it to be used intelligibly, the full gains will not be realized. For example, if a component coding system used in the warehouse is different to that used in design or production control activities, the potential gains from inter-sphere cannot be realized. Once again this requires a recognition by management that production is a fully integrated process made up of individual, harmonized subsets, rather than a series of separate if vaguely linked activities.

Associated with this is the rationalization of design and process to limit the number and types of components, hence allowing this harmonization of information bases to proceed in a productive manner. The central concept here is that of group technology (GT) which was initially developed in the USSR and Germany in the 1950s, and subsequently sank into obscurity until the emergence of intra-sphere and inter-sphere automation technologies in the late 1970s. Although there are a variety of different definitions of this concept the central idea is that enunciated by Salaja:

> Group-Technology is the realization that many problems are similar and that, by grouping together similar problems, a single solution can be found to a set of problems, thus saving time and effort (quoted in Halevi, p77).

Two examples may be given of group technology and the potential gains it offers. The first is that arising in design where the recognition that there exist 'families-of-parts' has allowed firms to reorganize designs into a more modular form taking advantage of basic component designs. In the pre-CAD days when designs were all manually done, it was not necessary to utilize 'families-of-parts' methodologies, but the capability of CAD to store basic designs and to scale them up or down automatically has vastly increased the potential benefits to be gained from this form of group technology. A second example is that arising from workcells: Halevi cites studies which estimate that using conventional technologies only around 5 per cent of production involves direct working; of this only 30 per cent involves machining and 70 per cent is taken up with positioning and tool-changing. Thus only around 1½ per cent of total manufacturing time is taken up by actual machining. Workcell concepts are thus being developed to change this rate by grouping together similar types of process and parts and to consequently reduce the non-machining component in production. Group technology is still in an embryonic form; moreover it will probably never be possible to entirely automate this classification system (Halevi, p55). But, nevertheless, the systematization of information regarding design and process, and its links to automation technologies, is undoubtably an important factor influencing both the time scale and the extent to which the benefits of inter-sphere automation will be realized.

A TIME SCALE FOR INTER-SPHERE AUTOMATION

Given the various sets of impediments discussed above it is not easy to determine the time scale within which inter-sphere automation will emerge. In particular, as we have pointed out, these obstacles probably have more to do with political and social factors (such as the relative power of labour and capital, particularly in the context of individual countries) than technological ones, and are hence difficult to predict. However, a number of observations can be made. First, we have already begun to see examples of inter-sphere automation and these technological link-ups are occurring at an extremely rapid pace. Second, many of these new technologies are being innovated in an incremental manner and, despite the emergence of automation-facilitating technologies such as local area networks, this piecemeal approach may well hinder the pace at which full inter-sphere automation will proceed. Third, many of the more popularly celebrated examples of the automated factory (such as the Fujitsu Fanuc factory in Japan which employs 100 people on the day shift and only one at night) are really exercises in intra-sphere automation, largely confined to the manufacturing sphere. And fourth, it is in the less glamorous context of CAD that the first real signs of inter-sphere automation are beginning to emerge.

We are therefore faced with a picture of incremental technological change. But we should not confuse this incremental, piecemeal approach with the pace at which it proceeds and here the signs are that the speed of diffusion of inter-sphere automation is rapid. Once the coordination sphere sees widespread intra-sphere automation (a phenomenon which is likely to occur within the 1980s) then we can anticipate the emergence of full inter-sphere automation, the true 'factory of the future'. In summary therefore the 1980s are likely to see increasing signs of individual sets of inter-sphere automation technologies and by the 1990s we will surely see the fairly widespread emergence of fully automated production in many sectors including those now characterized by small batch production.

Lest this be seen as fanciful or insufficiently cautious it is instructive to review the progress of intra-sphere automation in the manufacturing sphere since Bright (1958) wrote his classic study in the mid 1950s. As we saw in Chapter 2 Bright foresaw 17 different levels of mechanization (see Figure 2.3); however, in the 13 advanced plants which he monitored, mechanization seldom exceeded level 6. In 1958 Bright had this to say of the prospects of mechanizing the remaining levels:

> It should not be inferred that all operations will therefore continue to rise and eventually reach the 17th level (Anticipatory Control) or something close to it. Quite the contrary – although there is a strong economic spur to raise activities off level 1, 2, and 3 (even 4), there appears to be an equally strong economic rein after reaching levels 5 and 6. In many cases there literally is neither necessity nor economic advantage in achieving mechanization above, say, level 6, with an occasional use of levels 8, 9 and 10 where needed (p223).

Now Bright was observing a world which was largely in the pre-electronic era; in the absence of the flexibility which electronics allows (as in the flexible manufacturing systems discussed in Chapter 4) Bright was probably correct to anticipate obstacles to more extended mechanization. But the introduction of electronics has changed all this and, barely twenty-five years after, Bright made these observations:

> Todays advanced CNC machine tools and 2nd generation robots achieve approximately level 16 in Bright's classification (Husband, 1982 p2).

Similarly when referring to Amber and Amber's ten orders of automation (Figure 2.2), which are more ambitious then Bright's levels of mechanization, Husband observes that

> Todays 1st and 2nd generation robots start to meet A6 [ie Order 6] characteristics (passim).

With the benefit of hindsight, therefore, it is wise not to underestimate the extent of possible inter-sphere automation. Moreover, given the effect of the growing crisis in the world economy on competition between firms, and their technological response to this competition (see Chapter 1), it would similarly be foolish to underestimate the pace at which this automation will proceed.

PART THREE

The Impact of Automation

Earlier chapters have concentrated on explaining the contemporary drive to automation, in categorizing the different types of automation technologies available and in mapping their diffusion into the three spheres of production. But although this tells us about the origins, the nature and diffusion of automation technology, it has had little to say about its impact. Given that we are witnessing significant changes in technology and in the organization of production, it is important to be aware of these effects. Therefore, in the three chapters which follow we examine the likely impact of automation technologies on capital and labour in developed countries and on the division of labour between developed and underdeveloped economies.

However, before we proceed with this analysis, two sets of caveats are in order. First, these three sets of actors are not homogenous groups. Capital, as we saw in Chapter 1, comprises of different fractions such as 'national' and 'transnational' groupings; similarly labour is made up of different elements reflecting factors such as skills and union-membership; and the Third World not only comprises of an heterogenous set of countries at different stages of industrial development, but also of different classes which often hold antagonistic interests. To some extent these complexities will be specifically recognized in the discussion which follows, but, inevitably, at some point, we will be forced to collapse these differences and refer to 'capital', 'labour' and 'developing countries' as if they were homogenous entities.

The second caveat concerns the questions of whether it is feasible to discuss the 'impact' of a technology as a thing in itself. As we shall see in Chapter 8, technological progress does not follow immutable laws which govern its precise nature. On the contrary, it is subject to manipulation and is more often the conscious implement of particular groups, best understood in relation to the struggle for power within the enterprise itself. Consequently, in discussing the impact of automation technologies on capital, labour and Third World, it is best to see it as an association between particular types of technological developments and the emerging nature of the social organization within which these technologies are introduced. The term 'impact' is used as a type of shorthand for this dialectical interplay between technology and society and this will become apparent as the discussion proceeds.

CHAPTER SEVEN

The Impact upon Capital

In the fantastic world of full automation in which all labour is incorporated in a self-generating, self-correcting and self-operating machine (often called 'dead labour') it would be possible to talk of capital as a thing in itself. However, since these conditions do not apply (nor could they) it is necessary to see capital's position as being defined in relation to labour, as well as to the state and other fractions of capital. Thus, capital can be defined as that constellation of forces which owns machinery, buys in labour and other inputs, organizes production and marketing, is responsible for production and process development, liaises with the state and copes with competition from other firms. As such, in order to operate effectively, it is essential that it retains control over social, technological and economic relations of production. So for capital, the role of technology is not only to produce attractive commodities at low cost, but also, as we shall see in the following chapter, to allow control to be exercised within the factory itself.

In this chapter we will examine the links between emerging automation technologies and the balance between different fractions of capital. Two different but overlapping sets of fractions will be considered: differently sized firms and national versus international firms. It is important also to distinguish between firms producing automation technologies and firms using automation technologies. But before we can move on to this detailed discussion of the relationship between automation technologies and size and type of firm, it is useful to consider briefly those factors which lead to the concentration of production into limited numbers of firms. We shall thus be able to conceptualize different types of capital, rather than seeing all firms as atomistic and undifferentiated entities.

THE CONCENTRATION OF OWNERSHIP AND PRODUCTION

This is a complex subject, treated extensively in a wide variety of empirical and theoretical studies. Because we have only limited aims in this chapter, we will not attempt to review the literature on industrial structure but rather will draw out some important elements which are relevant in the discussion we are pursuing. (Readers interested in pursuing this matter in greater detail might begin by consulting, Bain, 1956, Pratten, 1971, Brett, 1983 and Blair, 1972.) The first point to note is the tendency towards growing concentration of ownership of production. For example, in the UK the proportion of total manufacturing net output accounted for by the hundred largest enterprises rose from 16 per cent in 1909 to 24 per cent in 1935 to 32 per cent in 1958 and to just over 40 per cent in 1970; concentration in the USA rose in a similar manner over this period (although to a lesser extent), that is from 22 per cent in 1909 to 33 per cent in 1970 (Stewart, 1978). Second, accompanying this trend towards concentration of ownership has been a constant tendency towards the concentration of production and hence unevenness in industrial development. This disparity between regions has taken both national and international forms and is particularly evident in relation to the Third World. Thus despite possessing around 75 per cent of the world's population, developing countries currently only account for less than 10 per cent of global manufacturing value-added. This share has tended, from the mid-1960s to grow (as we shall see in Chapter 9); but the question is whether continued growth of this share is likely in the 1980s. (Value-added here refers literally to the value added in production, that is the difference between the value of the output and the cost of raw materials and components.)

A variety of factors explain this trend towards the concentration of production and ownership, a phenomenon which incidentally is not confined to the UK and the USA. Some of these can be classified as 'entry barriers', that is factors which make it difficult for new firms to enter the market such as minimum effective levels of scale of output (eg in mass-production industries), critical minimum R & D efforts and skill constraints. Other factors (often overlooked by economists who have had a tendency to focus disproportionately on entry barriers) are those reflecting the historical evolution and nature of particular capitalist classes and their links to the state and to financial capital.

However, since this book is primarily concerned with the links between automation technology and society, we will focus the discussion on particular types of entry barriers, rather than these wider political-economic factors, since the nature of technological progress over the years has been an important barrier to entry by new firms and has thus had a significant impact on the growing concentration of ownership and production. By doing so we shall be able to raise the question – rather than determine the answer, for our understanding of emerging automation technologies is still only at an embryonic stage – of whether these new technologies contribute to further concentration of ownership and unevenness of development.

Foremost amongst these technological factors underlying the concentration of ownership and production are what are called economies of scale. These economies are centrally related to the existence of fixed costs which are incurred irrespective of the rate of output. Consequently the higher the number of products produced, the lower the average cost, since the fixed costs are spread more widely. For example, as we can see in Table 7.1, there exist significant scale

Table 7.1 Variation in production cost in relation to different scales of output in selected industries

Product, capacity and cost	*Unit*	*Variation in capacity and production cost*			
Steel					
Capacity	Thousands of tons per year	50	250	500	1,000
Cost per ton	1948 U.S. dollars	209.4	158.8	137.5	127.2
Cement					
Capacity	Thousands of tons per year	100	450	900	1,800
Cost per ton	1959 U.S. dollars	26.0	19.8	16.4	13.9
Ammonium nitrate					
Capacity	Short tons per day	50	100	150	300
Cost per ton	1957 U.S. dollars	190.4	145.1	125.6	101.5
Beer bottles					
Capacity	No. of moulding machines	1	2	6	12
Cost per gross	1957 U.S. dollars	8.51	7.25	6.13	5.69
Glass container					
capacity	No. of moulding machines	1	2	6	12
Cost per gross	1957 U.S. dollars	8.66	7.77	6.78	6.33
Radial ball-bearings					
Capacity	Production index (1961=1)	1	2	3	
Cost per thousand	1961 yen	79,800	67,100	63,100	
Tar					
Capacity	Tons per day	100	200	300	400
Cost per ton	Thousand of 1961 yen	10.5	9.6	9.2	8.9
Benzol					
Capacity	Tons per day	50	100	200	300
Cost per ton	Thousands of 1961 yen	29.2	27.1	25.9	25.4
Aluminium plate					
Capacity	Tons per year	200	1,200	3,000	5,000
Cost per ton	Thousands of 1961 yen	276.8	272.2	269.1	263.5

Source: Stewart, 1978, p 62.

economies in a variety of important manufacturing sectors, and these conclusions are equally valid for a large number of other manufacturing sectors.

Once these scale economies exist there will be an inevitable tendency towards the concentration of production in large-scale, technically efficient plants, leading to the centralization of production in particular areas and in particular firms. But the precise nature of this concentration will reflect the factors which underlie these economies of scale and here it is important to distinguish between various causal factors. Since average costs of production (which, as we have seen, tend to fall with increasing scale) relate centrally to the total costs of production, it is important to distinguish between the two major elements of total production coasts, namely direct and indirect costs.

Direct production costs are those inputs directly involved in production such as machinery, labour, energy, raw materials, components and buildings. Of these, raw materials and components are generally used in a constant relation to the rate of output, that is any increase in output requires an equivalent increase in inputs. But the other major items of direct production costs do not generally vary in this constant relation to output and economies of scale do arise. Most significantly these scale economies are to be found in the use of machinery where engineers have come to use a particular rule of thumb, the so-called '0.6 rule' (see Chilton, 1960,) which states that the increase in the costs of equipment is generally in the same ratio (ie 0.6) to the increase in capacity as the relationship between surface area and volume. This means that the machinery costs of each unit of output decline with the scale of production. And since the technological history of the past three centuries has seen the constant tendency for machinery to be substituted for labour and hence unit machinery costs have become a dominant cost item, this has been a significant factor leading to economies of scale in production.

However, this '0.6 rule' tends to apply to the dimensional process industries described in Chapter 2. Industries producing discrete output (ie non-process industries) are likely to be affected by a second factor contributing to economies of scale in direct production costs, namely the 'costs of changeover'. By this we mean the costs involved in adjusting machinery to produce similar products, but of different specifications. These costs include not just the labour required to retool machinery but also (and often more significantly) the downtime during which machinery has to be stopped and production is lost. An example of these changeover costs, measured in terms of time and lost output is given for the glass container industry in Figure 7.1, in which the mould changeover costs (two hours downtime) are exacerbated by an initially high reject rate until the machinery specifications are adjusted on a trial and error basis to the optimum levels (this takes around six hours to get right). Hence, at least in some industries, high changeover costs are a substantial item and this is another major factor leading to the development over time of dedicated production lines producing a single size or type of product. The production runs required to avoid changeover are often so high that it leads to specialization between plants, firms, regions and countries.

There also exist a series of *indirect* production costs which lead to economies

Fig. 7.1 Rejection rate on changeovers in manufacture of glass containers

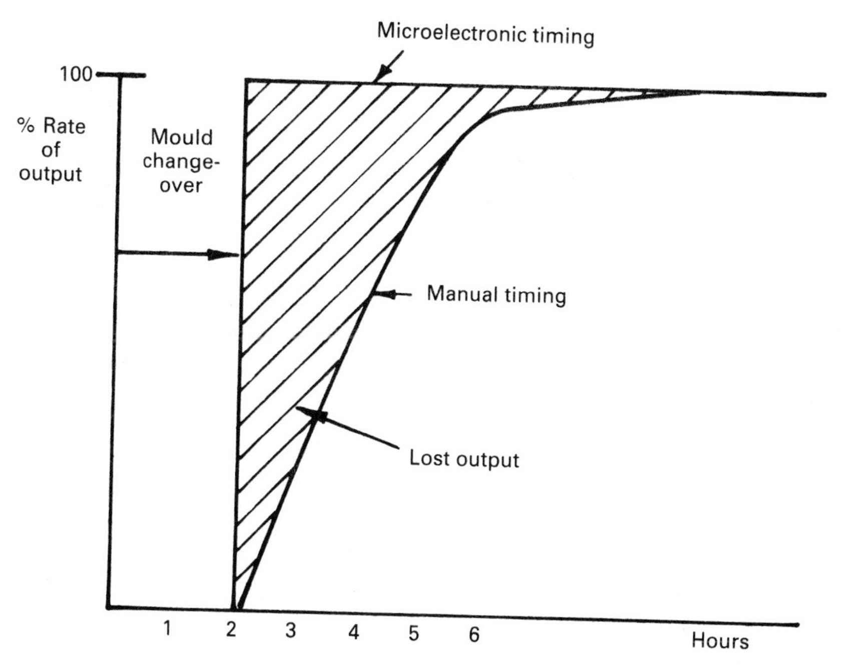

of scale. By indirect production costs we refer to activities which are a necessary background activity to production but do not in themselves feed directly into manufacture. Historically, four of these indirect production costs have stood out in significance, each involving economies of scale or effort. The first has been *R & D*, which as we saw in earlier chapters, has grown in importance as a specialized activity as the scientific content in manufacturing has increased. In general this R & D is associated with scale economies, for example, mastering the software for numerically controlled machine tools controls has allowed many firms to progress to the manufacture of transfer lines with little additional effort. But in addition, there are often minimum effective levels of effort in R & D, requiring the gathering together of a variety of complementary skills, which make it difficult for small firms to spread these R & D costs over limited production runs.

A second significant indirect production cost – but one which involves fewer economies of scale – is that of *management*. Certain functions of management – such as the establishment of various control procedures over stock or over labour – occur irrespective of the size of the firm. Increasing the number of these units under managerial control is often possible without any increase, or with a less than proportionate increase, in the scale of managerial effort. However, counterposed against these scale economies is the tendency for large, generally international firms to develop unwieldly and costly bureacratic structures. These often have the effect of muting the realization of the potential scale economies arising in management.

A third area in which scale economies occur in indirect costs is that involving the *purchasing of inputs*. The ability to buy in bulk, particularly in the case of non-standard items, often provides the opportunity for the component supplier to reduce its costs of production, hence lowering unit costs for the purchasing firm. In addition the bargaining power available to a firm which satisfies its global needs from a single or limited number of sources, is substantial, thus providing further economies to the large-scale producer. It is because of such benefits, for example, that most major motor manufacturers have been drawn to the strategy of 'global sourcing' and the development of the 'global car'.

Finally, scale economies inherent in *sales and marketing* are a further source of savings in indirect production costs. These economies arise from the ability to spread the costs of marketing – which are often fixed and with a minimum scale (for example, office, salesperson and secretary) – over a large number of items.

It is important to bear in mind that although these technology related factors do not tell us the whole story about concentration, they are an important (and often necessary) component of the explanatory factors. The value of this distinction between direct and indirect production costs is evident when we distinguish between the concentration of production in plants and in firms. In the former case, production is concentrated in particular locations to take advantage of scale economies, but these generally arise in relation to direct production costs such as those resulting from high fixed machinery costs or long changeover times. By contrast, when scale economies occur in indirect production costs, concentration is likely to occur in relation to firms, with production being spread between various plants, but R & D, management, purchasing and sales concentrated in regional centres.

Historically, the period until the 1930s was one in which there was a trend towards concentration with respect to plants. Thereafter, until the late 1960s, there had been a tendency for economies of scale in direct production costs to be relatively muted and the major growth in concentration occured in relation to firms, rather than plants (Brett, 1982). Then in the 1970s, as recent developments in the motor, aerospace and electronics industries illustrate, there was a renewed surge of concentration at the plant level reinforcing the trend towards firm concentration. The question now being asked is what impact the emerging automation technologies will have on the trend towards plant and firm concentration. It is to these issues which we now turn our attention.

The Likely Impact of Automation Technologies on Concentration

In examining the (rather scant) evidence on the relationship between the developing automation technologies and industrial concentration we begin by focusing at a general level. In doing so, we concentrate on the particular effect of electronics related automation technology, since as we saw in previous chapters, this key heartland technology lies at the root of the emerging developments in automation.

Thus with respect to scale economies arising in relation to direct production costs we can observe a mixed bag of likely effects. First, the '0.6 rule' or any other scale economy of this sort which tends to characterize industries producing dimensional products, is likely to be little affected. Whilst there is evidence that electronically controlled systems are often smaller and cheaper to construct and maintain, the costs of providing adequate software, especially as new systems are improved, means that the overall impact on scale economies of introducing these automation technologies is likely to be negligible. But the impact of of these automation technologies on reducing changeover costs, which occur in industries producing discrete products in sectors where a limited number of variants of a product are being made, will be substantial. For example, if we refer back to the glass container production line in Figure 7.1, we can see that the high costs of changeover in the pre-electronic era resulted from a combination of downtime (due to manual mould-changing and machine resetting) and production spoilage (until settings are adjusted). When electronic control systems are substituted for the previous system of electro-hydraulic and eletro-mechanical ones the resetting is instantaneous and exact; moreover, automatic tool changes assist in reducing downtime from over eight hours to less than one hour and in virtually eliminating spoilage. Similar stories can be told for a very wide variety of manufacturing sectors. Consider a second example of the Mazda car assembly plant described in Chapter 4. Until recently the production of each of the three different types of car (ie the 323 hatchback, the 626 rear-wheel drive and the RX7 sports car) would have been characterized by economies of scale, with average costs lowest at scales in excess of 500,000 cars per year. Now, because of the introduction of flexible assembly lines, the effective scale economies per model are much lower. In the case of the RX7, this may involve only a few thousand cars per annum, with the bulk of production concentrated on the more popular smaller and cheaper models. This would suggest that the erosion of dedicated production lines by flexible ones will have a major impact on reducing changeover costs and hence economies of scale in plants.

However, it should be remembered that flexibility always occurs in relation to particular constraints. For example, a flexible engine assembly line may be able to produce a number of variants of engines, but not able to produce gear boxes or assemble car bodies. Moreover, in many cases the enhanced flexibility is reflected in a plethora of differentiated products (such as the one hundred and eight different engines being produced by the new Fiat engine line described in Chapter 4) rather than in reductions in plant size. Nevertheless, although it is still at an early stage, there are signs that the new, flexible automation technologies may undermine plant economies of scale in direct production costs.

With respect to indirect production costs it is more difficult to foresee that the new automation technologies will reduce firm scale economies. For, as we shall see in later discussion, R & D in the new technologies is no less prone to scale economies; similarly scale economies in purchasing and marketing show little difference between the new and old automation technologies. In addition, the ability which informatics give management to limit 'control loss' (which, as we saw in Chapter 5, is a function of size) is likely to be a significant factor in increasing the tendency to economies of scale for the firm. This is particularly evident in relation to concentration across countries where international firms are rapidly developing an important capability in the sphere of cross-board data flows (United Nations, 1981).

We have considered the link between automation technologies and concentration at a general level. But how have factors worked themselves out at a detailed sectoral level, both in relation to the size and type of plants and firms involved? Unfortunately, since the new automation technologies are largely a recent phenomenom, adequate empirical information is difficult to obtain. With respect to the supplying industry we believe that an examination of trends in the interactive computer-aided design sector (hereafter CAD), supplying intra-sphere automation technologies to the design sphere of production, will be illuminating. For firms using new automation technologies relevant information is more limited and we shall be forced to rely on material which, whilst interesting, does not allow us to reach determinate conclusions.

Concentration in industries supplying automation technologies: The example of the CAD industry

In this section we will review the development and concentration of the CAD supplying industry and attempt to draw out those elements in its history which are likely to be analogous to other segments of industry supplying automation technology. *Market structure*: hitherto the market for CAD systems has been segmented into three major divisions, largely reflecting the distinction between graphics program (which only present visual images) and applications software (to undertake design analyses of varying degrees of complexity). The divisions are as follows. *Mainframe*-based systems are partly used for graphics and partly for information processing and analytical tasks; these involve large, established electronics firms with a wide range of products. *Minicomputer*-based systems are dedicated to graphics use. They can undertake a variety of different applications program, but with limited ability to perform complex analytical programs: these involve firms established in the early 1970s, almost all of which confine their products to various types of CAD systems. *Microcomputer*-based systems have a basic draughting software and are dedicated to a few applications programs with a very limited analytical capability; these comprise new, small firms, established in the late 1970s, with a single, particular CAD product.

The running so far has been made almost entirely by the minicomputer-based turnkey systems. The exceptions are the emerging sales of software systems produced by new, independent companies and IBM (whose presence has hitherto largely been limited to very large companies each using a relatively large number of graphics terminals with the processing mainframe also being used on a batch-basis for additional heavy analytic and data-processing activities). However, IBM only markets its mechanical software. Its electronic software is not marketed outside affiliates since IBM are concerned that this would divulge proprietary information regarding component, integrated circuit and computer architecture.

Currently seven US firms dominate this turnkey market. The most significant factor which emerged over the decade was the growing dominance of the market by a single firm (Computervision) whose market share grew from 28.1 per cent in 1976 to 33.2 per cent in 1980, at the expense of all of the other turnkey vendors. In assessing the evolving market structure in this industry producing automation equipment for the design sphere, it is possible to distinguish between four distinct periods in the industry's development.

In the initial pre-1969 period of industry development, the major actors were existing large firms in the defence, aerospace and aeronautical industries, collaborating with mainframe computer manufacturers such as IBM. The first technical breakthrough (the light pen and screen, allowing for interactive use) occurred in the 1950s, during the development of the SAGE early-warning radar system. By the 1960s, the aerospace industry became the major user, pushed by the US Air Force which actively attempted to widen applications to other manufacturing industries in the late 1960s. In summary, therefore, during this early period there was hardly any 'market' for CAD, with most developments occurring to assist own-use by large, technologically advanced mechanical-engineering corporations in the US and (to a lesser extent) in the UK.

In the short period between 1969 and 1974 the industry began to change its nature significantly by rapidly diffusing its products to the electronics sector. The primary impetus for this were new, small firms begun by software writers spinning off from other industries. These comprised two groups. The first were those with specific experience in CAD software in the aerospace and automobile sectors: a particularly influential group moved west to Southern California and formed a company called Systems Science Software which over the years provided the basic software for many of the current turnkey vendors; in another case a firm in Huntsville, Alabama, was established by ex-IBM employees taking up software writers displaced when the neighbouring Apollo moon-program was wound down. A second source of software writers emerged from the electronics sector itself (especially IBM) who were attracted by the low barriers to entry and the obvious future potential of the CAD sector. The consequence was a variety of new firms, initially making digitizing equipment and subsequently moving to the supply of complete turnkey systems. This includes all of the current major US turnkey vendors except IBM. By contrast, in Europe (and especially the UK) the emergent CAD capability in this period arose directly *within* established electronics firms (such as Racal, Plessey and ICL) who produced equipment for their own needs. In summary, therefore, this second period of industry development saw the emergence of new, independent firms and the rapid diffusion of the technology out of the defence aerospace and automobile sectors to the electronics industries.

From 1974 onwards the trend was towards concentration. By around 1974 most of the major suppliers were established and protected by a large investment of software in a suite of specialized applications programs. At this point CAD equipment began to penetrate manufacturing industry. The rate of diffusion was so rapid in this sector that the aggregate industry growth rate increased from around 55 per cent per annum to around 80 per cent per annum. All the vendors, despite historical specializations adopted a similar strategy of expanding their range of applications programs to provide comprehensive cover to all industries. But this diversification, together with the cost of financing such heady rates of expansion, necessitated the raising of financial resources. All firms – including those owned by venture capital – were forced both to sell shares in the stock market and into long- and short-term debt.

At the same time as these established CAD suppliers were raising funds, predators were emerging amongst established corporations. First, McDonnel Douglas took over United Computing in 1974 changing its name to Unigraphics. Then around 1975, Gerber, a major name in the garments sector, expanded its range of operations, buying in the basic graphics software. But most significant was the action of US-based General Electric (GE), which took over the second largest CAD firm, as part of its strategic decision to expand its automation technology activities (see Chapter 1). And, most recently, Schlumberger, the French multinational diversifying its activities from the narrow base of producing equipment for the petroleum industry into electronics, took over Applicon, one of the largest CAD supplying firms.

Therefore, one pressure towards concentration in this period was a vertical and horizontal integration resulting from enterprises in other sectors attempting to buy into a rapidly growing industry. A second tendency towards concentration was organic growth within the CAD sector itself, namely the growing market presence of Computervision at the expense of industry minors. Computervision not only increased its market share (from 28 per cent to 33 per cent between 1976 and 1980) but is currently becoming a transnational corporation itself with the decision in the late 1981 to set up a manufacturing subsidiary in Germany, and in 1982 to develop links with a firm in Singapore. To summarize, therefore, this third phase of industry development was associated with the growing size of CAD firms, the growing organic trends towards concentration within the sector, and a tendency for formerly independent CAD firms to be swallowed by existing transnational corporations.

In the most recent development which is only just emerging we can once see the emergence of new, small firms spun off from larger, older and established firms. To understand their significance we need to go a little deeper into the nature of CAD technology. CAD systems are built around two sets of software, basic graphics and applications. The advantage of mainframe-based systems are that they are not only able to cope with the heaviest requirements of particular applications programs, but they are also able to undertake complementary batch-processing tasks such as payroll and inventory controls. Their disadvantages are that they are costly (an entry cost of over $500,000 with unit-terminal costs of around $60,000), that they are vulnerable to breakdown of the centralized host computer and memory, and that the mainframe suppliers have hitherto developed only a limited range of applications programs. By contrast the minicomputer-based turnkey vendors have developed the most comprehensive range of applications software, most of which can be used independently from mainframe computers. They have been lower entry costs (around $200,000) with lower unit terminal costs. The major disadvantage is their weaker processing capability which constrains their use for heavy-analytical applications programs and hinders their response rate to users.

Recently, as the CAD vendors have penetrated a wider user base, a 'space' has begun to emerge for dedicated systems. That is amongst users (eg engineering firms using printed circuit boards and draughting firms) who have no need for a comprehensive suite of applications programs and whose engineering data-processing requirements are not extensive. Their needs can be met by microprocessor driven systems which provide basic graphics software with a single (or small number of) applications programs. Such systems, based upon already 'mature' applications programs developed by the turnkey vendors are rapidly beginning to emerge, selling for less than $30,000 each. Invariably these involve new firms begun by ex-employees of existing CAD vendors. To summarize, therefore, this most recent stage of industry development has seen two divergent trends: a continued tendency to concentration and an opposing tendency for the entry of new, small firms selling limited capability dedicated systems.

The key to the discussion of concentration in the CAD industry lies in the *barriers to new entrants*. For, given the high growth rate and profitability of the industry, we would expect the entry of many new firms to undermine any organic tendencies towards concentration of ownership and production. We, therefore, consider the problem of entry barriers in the CAD industry at some length because we believe the discussion throws some light on other emerging automation technologies.

Of the factors directly relevant to this discussion, the primary one protecting existing CAD suppliers from new entrants has been the scale of software inputs necessary to offer a competitive package of applications programs. In Table 7.2 we detail the extent of R & D inputs of these firms for which information exists; in Table 7.3 we detail (without mentioning the names of firms to avoid disclosing proprietary information) the current numbers of software writers they employ and, where available, the accumulated input of software person years in their system. It can be seen from Table 7.2 that compared to US industry in general, the CAD industry invests a very large proportion of sales in R & D. This proportion is high even relative to information processing in general where the 1979 average was only 6.1 per cent of sales and the 1980 average for all US firms was 2 per cent. From Table 7.3 it can be seen that most CAD suppliers currently employ over 100 software writers per year, expanding these numbers at well over 30 per cent pa. Some of the vendors were able to detail their stock of software – over 1,000 person years in some cases and over 7m lines of code in others.

These absolute sunken R & D expenditures are in themselves not a sufficient deterrent to entry. After all in 1980 prices even Computervision, the largest firm, had accumulated less than $80 million of R & D, and some of that was

Table 7.2 R & D expenditure as percentage of sales in CAD industry

	1969	*1970*	*1971*	*1972*	*1973*	*1974*	*1975*	*1976*	*1977*	*1978*	*1979*	*1980*
Applicon								11	8.4	12.8	8.9	11.2
Auto-trol[1]						22.8	13	14.6	12.6	15.7	14	12
Computervision	671	28	4.4	6.4	9.7	11.8	12.3	8.9	9.6	8.3	8.8	
Intergraph								6.5	5.7	7.6	10.1	11
Gerber												15.9
All US industry	2.2	2.2	2.1	2	2	1.9	1.9	1.9	1.9	1.9	1.9	2

Source: Kaplinsky 1982b

[1] In 1980, Auto-trol's ratio of R & D expenditure to sales (which at 12 per cent of sales, was the lowest ratio for the firm since 1974) was the fourth highest ratio of the 744 US Corporations surveyed by Business Week; its R & D expenditure per employee was the third highest of the sample.

Table 7.3 1980 software staff and accumulated person-years of software in eight individual firms*

Firm code	*Numbers employed*	*Accumulated years of software in system*
a	40	130 person years post bought-in package in 1974
b	150–200	1000 person years
c	103	No information (NI)
d	345	NI
e	125	NI
f	120	NI
g	110	600 person years in 2D draughting package plus 400 person years in bought-in mechanical outline
h	88	500 person years of software; 7 million lines of software.

Source: Kaplinsky 1982b.
(*) Excludes personnel on hardware development, but include those working on operating systems of minicomputers.

in the development of its own minicomputer; compare this with the $1.5 billion profits earned by General Electric in 1980 alone. The critical protective factors are that this software development occurs in a relatively specialized sector in the context of a general shortage of software writers. But most importantly, much of the necessary software development is sequential – as one of the founders of Computervision described it, 'One can't make a baby in one month with nine women.' A similar example can be drawn from the automobile industry: When Agnelli, the head of Fiat, was asked whether selling the Fiat 124 design to the Russians would not provide fatal competition to his firm, he answered, 'If we are still producing the Fiat 124 in five years time, we will go bankrupt anyhow!'

These barriers to entry are most substantial in relation to the comprehensive applications program systems offered by the existing turnkey suppliers using minicomputers and mainframe computers. There is, however, an emerging space for microprocessor driven systems which has been made possible by two factors. The first, and obvious, precondition, is the development of increasingly powerful microprocessors. The second is the maturity of specific applications programs which significantly lower the barriers to entry in these limited areas.

In order to grasp the significance of this latter point it is necessary to understand that in the CAD industry there is a distinction between the average and marginal cost of systems, which is the same as saying that there are economies of scale in the industry. The source of these scale economies lies in the imperative to all CAD suppliers to employ a large overhead of software writers. These software writers are in general divided into three groups. The first is committed to the development of systems which enable computers to operate, the second into a small number who upgrade and maintain existing applications programs and the larger number who are involved in developing the new applications programs which are necessary to keep an all-round presence in the industry. From each suppliers point of view, therefore, the software development process takes the form of a constant overhead investment in software, made up of a family of matured packages and an increasing family of new packages. The actual manufacture of CAD systems, by contrast, (which represent the marginal costs) is a relatively minor activity.

Microprocessor driven CAD systems are aimed at this market of matured applications programs, of which basic graphics capabilities (that is, the ability to draw lines, arcs and other forms which are necessary for computer-aided draughting) is the most obvious. From the vendor's point of view the software needs little attention and the system can be sold at a price which is close to its marginal cost, that is the cost of the hardware. There are a great number of such small firms springing up in North America and Europe. Many of them offer basic graphic capabilities with perhaps one applications program, often an auto-routing electronics program. Many of these firms are started by former employees of established turnkey suppliers who recognized that these turnkey firms were overstretched and unable to satisfy the limited needs of small-scale users.

We shall now examine *the relationship between the industry's growth curve and the trend to concentration*. The genesis and growing maturity of the CAD industry throws light on the pattern to be found in other industries supplying electronically controlled automation technologies involving a large component of software. In general the radical nature of the technology has meant that the origins of these industries are to be found in large user-firms who innovate to meet their own product needs. Thereafter, in the second phase, key individuals spin off from the pioneers to establish new, small firms dedicated to the development of the product. During the third phase these firms grow and become less vulnerable as accumulated software protects them from new entrants. At the same time their strong growth potential and their key strategic value attracts large predators from other sectors. In the final phase new, small entrants begun by people spinning off from existing firms find a role within particular mature applications programs.

Although no detailed studies have yet been conducted, similar patterns appear to be emerging in other automation technology sectors. For example, the word-processing sector appears to be rapidly moving into the third phase (trend to concentration), with new entrants from outside the subsector (eg Exxon) and organic growth from within (such as Wang). So too is the semiconductor industry in which many of the smaller independent firms are being taken over by existing, large firms (eg Philips takeover of Signetics in the USA) or becoming unviable (see Rada, 1982a). By contrast, the automatic testing equipment industry has already passed through this stage and is now seeing the emergence of small firms producing dedicated sets of equipment. Finally, in microcomputers, the entrance of IBM, Xerox and the Japanese firms appears to signal the transition between the second and third phases.

In many ways this notional growth pattern of software-intensive industries proximates that of the pre-microelectronic era. There is one crucial difference, however. In the earlier era, the mature phase was not characterized to the same extent by the emergence of new, small firms. This was because concentration was underpinned by plant economies of scale in direct production costs (eg automobiles). The software-intensive industries under discussion are distinctive because their relatively high input of software, with its associated variation between average and marginal costs of production, only allows for firm economies of scale in relation to indirect production costs. The manufacture of existing products (eg mature applications programs) is therefore open to new entrants if they can gain access to the technology itself. And in many cases, as we have seen in the CAD and automatic testing equipment industries, there are bits of technology (eg particular applications programs) which are often individual – rather than firm – specific, thereby providing the opportunity for new entrants.

Notwithstanding this confined 'space' for small firms in the mature phase of the industry's development, the tendency to concentration in industries supplying automation technology is likely to continue, despite the fact that the increasing flexibility of these technologies undermines economies of scale in direct production costs. However, unlike the 1970s when there appeared to be a parallel development towards increasing concentration at both the plant and the firm level, it would appear that in the production of automation technologies, plant and product economies of scale look likely to be undermined, while firm economies of scale, with its attendant tendency towards the concentration of ownership are likely to be enhanced. Finally, as we have seen, this concentration occurs not merely in relation to firm size but perhaps as significantly in relation to type of firm. Multinational firms become increasingly important actors, either because of vertical and horizontal integration, or because of organic growth within the sector itself.

Concentration in industries using automation technology

Unfortunately, there is only fragmented evidence to enable us to determine what types of firms are making first use of the new automation technologies. Four sets of this evidence give us a clue to the patterns of utilization which seem likely to emerge.

The first set of data comes from a telephone survey of twelve hundred manufacturing firms conducted in the UK (Northcutt et al, 1981a and 1982). The sample included representative firms from all the major manufacturing sectors, stratified into groups of two hundred enterprises covering the six major size categories, as measured in terms of employment. The survey was designed to determine whether firms were using electronics based technologies either in their products or their process (or both). Although the results are only presented by size groupings and therefore do not allow us to determine whether small firms are more or less likely to use electronics than large firms while producing the same output, (Table 7.4) they are nevertheless illuminating. They show clearly that the larger the firm, the greater the use of electronics, both in relation to products and process. Even when account is taken of the potential for using electronics (as perceived by the enterprises), it is clear that the largest

Table 7.4 Percentage of differently sized British firms using electronics in product or process

Use of electronics	Firm size – numbers employed						Average
	20–49	50–99	100–199	200–499	500–999	>1,000	
In process	13.5	20.5	38.5	46.5	64	85.5	45.1
In product	5	5.5	12	11.5	14.5	28	12.8
No potential use	54.5	38	24	14	7	2	23.3

Source: Calculated from Northcutt et al (1981a).

size grouping of firms tended to have taken up the use of electronics much more rapidly than their smaller counterparts. However, the very smallest enterprises (that is those employing less than 50 people) are relatively better than their medium-sized counterparts, perhaps reflecting the emergence of new, small, technology-intensive firms.

From this information it is clear that the use of electronics in process (which, ex hypothesis, reflects the use of automation technologies) is considerably greater by large firms than small firms, even when account is taken of the lack of potential for use. (However, it is worth noting that the enterprises were left to determine themselves whether the potential actually existed.) Although this pattern may exist because only large firms operated in these sectors in which automation could be used, this does not alter the primary conclusion that for whatever reason large firms seem to be taking up new automation technologies more rapidly than their small-scale counterparts.

The survey also tried to determine what type of firms made use of electronics either in product or process, distinguishing between independent firms confined to the UK, firms which were part of a large (probably multinational) UK-based group, and these enterprises which were subsidiaries of foreign firms. The results, presented in Table 7.5 are clear: UK based multinationals are more likely to use electronics in process (ie automation technologies) than independent UK-based firms, but much less likely to use automation technologies than subsidiaries of foreign firms.

The second set of evidence is drawn from a postal survey of 428 UK firms seeking to determine their use of computerized information systems for personnel management, that is automation technologies used in the coordination sphere of production. The evidence (see Table 7.6) confirms the conclusions of Northcutt and Rogers (who considered the utilization of automation technologies in the manufacturing sphere) that the larger the firm the greater the use of automation technologies.

A third set of evidence is drawn from a survey conducted by the Ministry of Labour in Japan concerning the diffusion of office-automation equipment. As can be seen from Table 7.7, there is a strong tendency for utilization to expand with firm size for seven types of intra-activity automation technologies.

Finally, it is possible to anticipate the diffusion of automation technologies by

Table 7.5 Percentage take-up of electronics by type of firm

	UK independent (503)	*UK group (609)*	*Subsidiary of foreign firm (88)*	*Average (Total 1200)*
In process	35	49	72	45
In product	9	15	22	13
In neither	62	46	24	51

Source: Calculated from Northcutt et al (1981a).

Table 7.6 Percent of British firms using computerized personnel information systems

	Blue collar employees	*White collar employees*
< 100	17.8	6.9
101–500	21.9	25.3
501–1,000	40.5	46.2
1,001–5,000	53.1	67.4
5001–12,000	67.6	70.4
> 12,000	63	76.2

Source: Institute of Manpower Studies

Table 7.7 Introduction of office automation equipment by size of Japanese firm (%)

Types of office Automation equipment	*Total*	*10,000-and-over employees*	*5,000–9,999 employees*	*2,000–4,999 employees*	*1,000–1,999 employees*
Total	100.0	100.0	100.0	100.0	100.0
General computers	87.1	94.3	94.1	89.3	79.7
On-line terminals	70.2	90.6	83.3	74.8	54.1
Office computers	36.1	58.5	44.1	31.6	30.9
Miscellaneous computers	30.8	56.6	35.3	31.1	21.7
Word processors	29.0	58.5	39.2	29.6	15.9
Facsimile equipment	60.2	84.9	68.2	57.8	52.2
Microfilm systems	34.0	62.3	51.0	36.9	15.5

Source: Japan Foreign Press Center, 1982.

asking the suppliers of these technologies what determines their marketing strategy and the pattern of their user network. When asked the former question, the vice-president of one of the largest CAD suppliers replied:

We use the DEC [ie Digital Equipment Corporation, the largest and most successful of the minicomputer firms] strategy: start at the top of the Fortune 500 list and work our way downwards. [*Fortune*, a major US business journal, issues an annual list of the 500 largest corporations in the USA.]

This is a clear statement of the self-reinforcing interaction between market size and marketing strategy, being reflected not only in the tendency for large firms to use automation technologies first, but also – because size and multinationality are closely correlated – for the international fraction of capital to gain a major technological advantage over their national counterparts.

A further element affecting regional concentration is the fact that there are substantial economies for producers of automation equipment when customers are located in the same area, since software and hardware support-staff each service a number of customers. Consequently, equipment suppliers place greater emphasis on marketing in areas where they already have a customer base, thereby further enhancing regional concentration. This has implications for the users as well, since, particularly in evolving software-intensive sectors such as CAD, users' groups have an important role to play in cross-fertilizing each other and in exerting pressures on equipment suppliers to maintain adequate support services. Hence users operating automation technologies in the same region are likely to make more effective use of their equipment, thereby further contributing to the existing uneven pattern of industrial development.

In conclusion, therefore, this chapter has examined the question of whether the new automation technologies are likely to be associated with greater or lesser degrees of concentration in production and ownership. Although we recognize that the concentration of capital is not primarily caused by technological factors, we were able to reach two major conclusions. With respect to plant concentration it would appear that the increased flexibility of electronics-based automation systems acts to undermine centralization. However the greater software requirements of electronics-based technologies coupled with the ability they provide to command information, are likely to be associated with greater firm concentration. Three sets of evidence are available in this regard and all point to the association of the new electronics technologies with large and international firms, rather than their small and national counterparts.

CHAPTER EIGHT

The Impact on Labour

In this chapter we examine the impact of automation technology on labour. But in the same way that we noted in the previous chapter that 'capital' could not be isolated as an independent entity outside its relationship with labour, so it is necessary to understand that the impact of automation technology on labour has to be understood through an examination of its links with capital. For, as we observed at the beginning of Chapter 5, one of the primary functions of management – as the representative of capital – is to control the labour force; technology, as we shall see, has an important role to play in the exercise of this control.

This chapter treats two subject areas. First we examine, at a general level, the relationship between technology and social relations with a view to situating the role of automation technology and explaining why it has taken a particular path of development. As part of this analysis we review the evidence with respect to the impact of automation on skills. Second, we conclude with a discussion of the likely effect of automation technology on the level of employment, referring once again to our earlier theme of the link between crisis and automation.

TECHNOLOGY AND SOCIAL RELATIONS: THE IMPACT OF AUTOMATION ON THE NATURE OF WORK

The relationship between technology and social relations is a complex one and has become an increasingly important area of investigation and debate. Most observers would agree that there is a direct link between technology and the organization of society. For example, large-scale machine-paced production lines to some extent determine the degree of centralization in society, and the nature of work and social relations at the point of production. But there is less agreement about the impact of social relations on technology – do particular types of societies produce particular types of technology? It is to this latter issue which we direct our attention, with the idea of understanding the particular configurations which the emerging automation technologies appear to be taking. There are two extremes of discussion supplemented by the usual attempts to offer a more 'refined and complex' theory which steps outside of the restricted bounds of either/or categorization. We begin with a brief examination of the two extremities.

The argument that technology follows its own logical path of development irrespective of the social context in which it is generated is usually conducted at an implicit level. Technology and technical change is seen as unproblematic, something which arises as manna from heaven. Hence the literature on production function in economics, for example, tends to confine itself to the question of whether there are constant, increasing or decreasing levels of scale. More recently, though, economists have come to recognize the crucial role which technology has played in increasing production. One celebrated study of the sources of economic growth in the USA (Denison, 1962) argued that less than 30 per cent of output growth was explained by an increase in the input of land, labour and capital; the rest was explained by new technology and organizational practices. Attention has therefore come to be focused on the sources of technical change, the rate of technical change and the factors inducing technical change.

Sometimes attention is also given to the type of technical change, but here the analysis is confined to 'economic' factors such as the degree of capital-intensity, the scale of output and so on. Consider, for example, one well-known textbook on economic growth (Eltis, 1966) which devotes a whole chapter to 'Research and Technical Progress'. In this Eltis distinguishes between two types of technical progress, namely embodied ('. . . technical progress which would not be exploited without new [physical] capital' p35) and disembodied ('. . . which could be exploited without new [physical] capital' passim). With respect to the former Eltis writes:

> A great many factors affect the rate of technical progress. Of these, many are likely to be sociological because education, technical mindedness, and willingness to accept change probably influence technical progress (p35).
>
> Industrial research and development must make a considerable contribution to t' [embodied technical progress] . . . The incentive to spend money on research and development is likely to depend on many things of which the amount of competitive pressure to which firms are subject, and the size and security of their markets might be among the more important (p36).

Here in this discussion of embodied technical progress we see no recognition of the nature of technical changes with respect to the social relations involved. This is perhaps not surprising, but it is significant that even in the discussion of disembodied technical progress, Eltis also fails to open up the technological black box.

> There is little that can be said about the factors which would be likely to stimulate 'disembodied' technical progress . . . Such improvements might be discovered and implemented by workers or managements and it is clearly important that they should be adequately rewarded (p42).

This approach – which is echoed in the overwhelming majority of economic texts on technology and technical change – implicitly assumes that technology is something to be 'discovered' and that its precise nature, with respect to the detailed work practices involved, is unproblematic. By omission it is assumed that there is no set of alternative work practices available.

At the other extreme are observers who start from a very different perspective. For them the factor to be explained is not so much the aggregate level of output, but the type of social relations in which production takes place. In capitalism, the fundamental contradiction arises between capital and labour and hence technology is seen as reflecting the imperatives of this struggle. We shall examine these ideas in a little more detail since the evidence now emerging in relation to automation technologies plays a key role in resolving the dispute between these two extremities of analysis.

The discussion begins with Smith, Marx, Babbage and Taylor. Marx helps us with two major ideas. The first is that of the key distinction between conception and execution in work.

> A spider conducts operations that resemble those of many a weaver, and a bee puts to shame many an architect on the construction of her cells. *But what distinguishes the worst architect from the best of bees is this, that the architect raises his structure in imagination before he erects it in reality*. At the end of every labour-process, we get a result that already existed in the imagination of the labourer at its commencement [emphasis added] (op cit, p178).

Marx then argues that the division of responsibility between these tasks reflects the social relations in which production takes place.

Secondly, Marx makes the distinction between the formal and the real subordination of labour. In the former the labour process is largely controlled by the labourer, whereas in the latter the capitalist has assumed control over its organization and the nature of work. It is interesting to note that there has been a tendency in Marxism to assume that this transition from formal to real subordination is largely complete in modern-day capitalism. However, insofar as formal subordination tends to be associated with craft-based work in which the labourer controls the pace of his/her work, it is important to re-emphasize the observations in Chapters 4 and 6, that around 75 per cent of output in manufacturing industry occurs in batches of less than one hundred. Consequently most of this work is not subject to mechanization; it remains skill-intensive and labour-paced. Hence it would be more realistic to talk of a constant *tendency* to real subordination – which as we shall see, automation technology reinforces – rather than the existence of real subordination as a *fait accompli*.

Smith, Babbage and Taylor contribute to the discussion by raising the issue of the division of labour. Smith argued that there were three factors lying behind the emergence of divided tasks: the existence of specialized machine builders, the saving of time if workers did not change tasks and the development of dexterity (ie skills). But Smith implied that the increasing division of labour was driven by the desire for greater output. He did not reflect on the social relations within which production took place. It was Babbage (who incidentally invented the first computer) who first drew attention in 1832 to the link between technological progress, the division of labour and *capitalist* social relations. He pointed out that it was the pursuit of profit (obtained in this case by sorting out skilled from unskilled tasks and thus employing cheaper – not less – labour) which underlay the division of labour which was emerging:

> . . . the master manufacturer, by dividing the work to be expected into different processes, each requiring different degrees of skill and force, can purchase exactly that precise quantity of both which is necessary for each process; whereas if the whole work were executed by one workman, that person must possess sufficient skill to perform the most difficult, and sufficient strength to execute the most laborious of the operations into which the task is divided (Babbage, 1832, pp137–8).

This procedure of organizing the labour process to reduce the *costs* of labour – which is not necessarily equivalent to reducing the *input* of labour (ie the so-called Babbage principle) – became more systematic with the development of the scientific management school in the late nineteenth century. F. W. Taylor first formalized this body of thought in his two major books *Shop Management* (1903) and *Principles of Scientific Management* (1911). As Braverman (1974) in his important study of the labour process points out, there were three basic principles which are to be found in the collective works of Taylor. The first concerns the 'dissociation of the labour process from the skills of the worker' (Braverman, p113), which Taylor describes in the following way

> The managers assume . . . the burden of gathering together all the traditional knowledge which in the past has been possessed by the workman and then classifying, tabulating and reducing this knowledge to rules, laws and formulae (Taylor, 1911, p36).

Second is the separation of conception from execution or in Taylor's words:

> All possible brain work should be removed from the shop and centred in the planning and laying out department (Taylor, 1903, pp98–9).

And third, management specifies in detail the tasks of each labourer.

The mechanism used to achieve these ends was the development of eight layers of functional foremen, which developed into the lower stratum of management discussed in Chapter 5. But the object of this exercise was unabashedly to increase profit:

> The adoption of standard tools, appliances, and methods throughout the shop, the planning done in the planning room and detailed instructions sent from this department . . . permit the use of comparatively cheap men even in complicated work (Taylor, 1903, p105).

To summarize this latter set of views we quote from Braverman whose book reopened much of the debate with respect to technology and social relations.

> Machinery offers to management the opportunity to do by wholly mechanical means that which it had previously attempted to do by organisational and disciplinary means. The fact that many machines may be paced and controlled according to centralised

decisions, and that these controls may thus be in the hands of management, removed from the site of production to the office – these technical possibilities are of just as great interest to management as the fact that the machine multiplies the productivity of labour (Braverman, 1974, p195).

So far we have contrasted two perspectives against each other. One sees technology as being unproblematic, following certain 'natural paths' of development: the technology is neutral and bears no significant relationship to the social relations in which it was conceived or to the work practices it involves. The other view argues that technology must be seen in the context of its development, notably as an instrument whereby capital controls labour, and that it consequently carries with it four important types of social relations. It involves hierarchical work patterns, is associated with increasingly fragmented tasks (the Babbage principle), it deskills the work, and it separates the conception of work from its execution.

So much for discussion at a general level. But what is the evidence for the relationship between new automation technologies and social relations? Here, once again, we have two extremes of argument. The first, general and largely unsubstantiated set of views, argues that automation technology is associated with an increase in the quality of work, a reduction in alienation and an enhancement of skill-requirements. Perhaps the most well-known exponent is the industrial sociologist Blauner, although it is not accurate to see him as purely general in approach since he draws empirical examples from four industries, printing, textiles, automobile assembly and chemicals. In his classic study *Alienation and Freedom* (1964) Blauner distinguishes four dimensions of alienation: powerlessness, meaninglessness, social alienation and self-estrangement. Whilst accepting that these result from the technology used, Blauner argues that not only is this technology autonomously determined, but that a U-curve of alienation exists ultimately resulting in increased job satisfaction and decreased alienation. Four types of technology are postulated – craft, machine-minding, assembly lines and continuous process (ie automation) – and these reflect a period of historical evolution. Whilst the transition from craft to machine minding and then to assembly lines is characterized by increased alienation, the move to the final stage of automation shows a decline in alienation. Thus, for Blauner at any rate, technology is not seen as an instrument of control and automation means that routine tasks are absorbed into machinery, a large repertoire of more sophisticated skills are required, and this involves a greater degree of autonomy and initiative by the workforce.

The evidence however, particularly that which examines the role which electronics-based automation technologies have to play, points to the contrary conclusion. Namely, that automation as we know it continues the trend in the capitalist labour process towards increased job-fragmentation, deskilling and greater control by management as the representative of capital; and that technological progress, as represented by the emerging automation technologies, is not immutable: it reflects the power relation within which it is generated and introduced. We will discuss a variety of evidence to illustrate this argument, treating the illustrations chronologically.

The first of the contemporary studies examining the links between automation and the nature of work is the classic study by Bright which we referred to in our discussion of the manufacture sphere in Chapter 4. Bright, working from the perspective of the Harvard Business School, concluded that if

> this machine evolution leads to the very situation feared: Automaticity grows until *no* operations are required: then what employment opportunities exist in the factory? Only maintenance, set up, and design. The level of education required on some of these design jobs is of Ph.D caliber. The maintenance skills require electronics training [a scarce factor in the 1950s when Bright wrote.] Now the unskilled and semiskilled worker is helpless. He does not even understand the language, let alone the job. Automation, if this is true, ultimately will raise the skill requirements for employment by eliminating ordinary jobs (Bright, 1958, p188).

However, Bright goes on to argue, a number of counteracting factors intervene. First, even maintenance, design and set-up tasks are being automated. Second, not all automation technologies are economically feasible even though they may be technologically so. Third, industry does not involve as much mass production as is commonly believed. And finally, machinery manufacturers consciously strive to downgrade the necessary operator skills. Given, as we have seen in earlier chapters, that new automation technologies are cheaper and more flexible, and that this cuts some of the ground away from Bright's second and third reservations, it is clear that his observations would seem to bear out the argument which we are pursuing.

Braverman concentrates on the development of numerical control to illustrate his belief that these automation technologies reflect the continued degradation and deskilling of work. He observes that there is no necessary reason why, with numerical control, the processes of conception and execution should be separated, but observes that:

> the unity of this process in the hands of the skilled machinist is perfectly feasible and indeed has much to recommend it, since the knowledge of metal-cutting practices which is required for programming is already mastered by the machinist. Thus there is no question that from a practical standpoint there is nothing to prevent the machinery process under numerical control from remaining the province of the total craftsman. That this almost never happens is due, of course, to the opportunities the process offers for the destruction of craft and the cheapening of the resulting process of labour into which it is broken . . . (p199).
>
> Each of these workers is required to know and understand not *more* than did the single person of before but *less*. The skilled machinist is, by this innovation, deliberately rendered as obsolete as the glassblower or Morse code telegrapher, and as a rule is replaced by . . . operatives (p200).

Braverman also draws on the conclusion of Bright that

> the relationship of skill requirements to the degree of automaticity is a declining rather than an increasing ratio (p215).

Noble, 1976, carries the numerical control case study a significant step further in illustrating the relationship between automation technology and social relations. He goes back to the early 1950s and examines the early develment of numerical control and counterposes it with an alternative technology, record playback. This latter technology was relatively trouble-free and mature in the 1950s, whereas numerical control took at least another twenty-five years to become a practical proposition for everyday manufacturing. However, the crucial difference between the two technologies was that record playback was built upon the skills of the machinist since it used these skills to 'learn' the task confronting it. By contrast, numerical control depended upon defining an abstract toolpath in mathematical terms and was a task executed by designers and engineers in the office. The brunt of Noble's argument is that numerical control was developed (with considerable federal support, especially from the military) at the expense of record playback precisely because it took control away from the shopfloor. Consequently, the development of numerical control reflects the desire by capital for control and is deliberately used to undercut the skill of the machine operator, thereby not only cheapening the cost of labour, but also degrading its content and the resultant job satisfaction.

The same point is taken up in a recent study by Shaiken (1981) which, perhaps because it is more recent, more explicitly recognizes the significance of electronics in allowing greater flexibility. Precisely because of this flexibility it undercuts much of the basis for skilled labour-intensive batch production (see Chapters 4, 6 and 7) and erodes the craft-skills in manufacture. Shaiken provides a great many examples to illustrate this point, perhaps one of the most graphic being the following statement by a machinist who had recently been assigned to a numerically controlled machine:

> I've worked at this trade for seventeen years. The knowledge is still in my head, the skill is still in my hands, but there is no use for either one now. I go home and I feel frustrated, like I haven't done anything (p29).

The second major attribute of the flexibility of new electronics based automation technologies is that for the first time it introduces machine-pacing into craft-skills which were formally labour-paced. A good example of this is draughting (Kaplinsky, 1982b) where a number of respondents argued that CAD completely changed the nature of their work. One draughtsperson observed that in the old days his work was craft-based – he illustrated this by proudly presenting an example of lettering. It was also social – he illustrated this by picking up an old drawing in which he had pencilled reminders of a dinner date and a technical problem raised by a colleague, and estimated that formerly he physically drew for only around 30 per cent of his day. However, the advent of CAD had radically degraded and deskilled his working life. It was no longer social – he now sat in front of a computer screen the whole day; his skills were no longer special – all users of the CAD system produced identical quality lettering (although, as he sarcastically pointed out, he had 63 different, automated, font-styles to choose from!); his work was much more intense – the CAD screen now paced his work with 'user-friendly' promptings and he had to work night shifts since the capital-intensive CAD equipment needed to be fully utilized to justify its purchase. This example is not unique to the CAD industry; Shaiken gives a number of examples drawn from other activities in other industries to illustrate that the new, flexible electronics-based automation technologies enhance the trend from formal to real subordination.

The final set of evidence we cite is that of Wilkinson (1981) which gives a slightly different flavour. Wilkinson, whilst agreeing that the nature of technology ultimately reflects (and, in turn, reinforces) the pattern of social relations, offers a more complex picture than the binary capital/labour antagonism considered by most observers. He researched in twelve different UK enterprises introducing electronics-based automation technologies and confirms the view that the technology is indeed flexible and that it reflects the drive for control. However, he opens up the capital/labour box to show that each is composed of different groups; there are different layers of management and different types of labour. In each of the plants visited the precise form which the technology took reflected the locus of control in the enterprise. (The same point is often made by some critics of Braverman – eg Elger (1979) and Friedman (1977) – that he ignores the implications of labours' resistance to the capitalist labour process.) In the light of his evidence Wilkinson concludes from his study of the introduction of new electronics-based automation technologies:

In fact, what our case studies show is that arguments about the efficiency of new production technologies are often *no more than scientific glosses which conceal or obscure the potential considerations* which have gone into decisions on technical change and work organisation. In other words, arguments from efficiency are used by the various interest groups in order to *justify, or make legitimate choices which are essentially political*, both in motivation and consequence (Wilkinson, 1981, pp194–5).

In conclusion, therefore, we have examined the evidence with respect to the relationship between automation technologies and social relations and have concluded that, contrary to Blauner, the path of technological progress is not fixed and the development of automation technology does not reverse previous patterns. These conclusions are underlined by a final quote drawn from the journal of American mechanical engineers (*Iron Age*):

Numerical control is more than a means of controlling a machine. It is a system, a method of manufacturing. It embodies much of what the father of scientific management, Frederick Winslow Taylor, sought back in 1880 when he began his investigations into the art of cutting metal. 'Our original objective' Mr Taylor, wrote, 'was that of taking the control of the machine shop out of the hands of the many workmen and placing it completely in the hands of management' (*Iron Age*, 30 August 1976, p156).

Before concluding this discussion of the sorts of labour processes involved in the new, flexible, electronics-based automation technologies it is necessary to expand briefly on a point touched on in our quotation from Bright. Namely, granted that automation technologies involve increasing job-fragmentation and deskilling of established tasks, but do they not at the same time create a requirement for new skills such as software programming and maintenance? This discussion is complicated since we are still at a relatively early stage in the development of these automation technologies and it is not yet entirely clear which way the technology will mature. Indeed, by the very nature of our above discussion, the final picture cannot be pre-determined since technology is still 'open' and through struggle it is possible (though not necessarily probable) that new technologies may involve radically new work structures.

Despite these caveats it is still possible to make two observations. The first is that it seems unlikely that new skills will be required in sufficient numbers to compensate for the displaced ones. This was in fact one of Bright's major observations and we shall examine some of the relevant evidence in the second half of this chapter. Second, many of the new skills and their attendant technologies are subject to exactly the same labour processes as the ones they replace. Take software as an example (Kraft, 1977). Here what was initially a highly skilled craft has increasingly come to be a routinized task. 'Structured programming' (that is programming according to house rules) has been introduced to lessen the dependence of an enterprise on the skills of any one worker. Programming tasks have been allocated in such a way that few individuals have an overview of the whole problem: for example, in one of the largest CAD supplying firms with a turnover approaching $100m, only three people have access to the full suite of software programs and tasks have been so arranged that few employees have an overview of the applications programs being developed (the primary aim is to reduce the vulnerability of the firm to the departure of individual software writers). Furthermore, software sub-routines are being automated and being embedded in hardware ('firmware' in the jargon) in order to increase the productivity of programmers and hide programs from competitors.

A similar series of events is occurring in the area of repair and maintenance in which the existance of duplicate systems (available due to the cheapness of electronic hardware) reduces the degree of breakdown. (These are called 'self-healing' systems in the industry.) In addition, self-diagnosing systems in which the 'engineer' merely slides out one printed circuit board and replaces it from stock, bypass the need for the worker to understand which circuit has broken down and why. The nature and scale of these skills are substantially lower than those involved in the repair and maintenance of pre-electronic mechanical control systems. Although there is no doubt that over the long run there is an increase in the 'average' skills of the labour force, the emerging skill requirements of automation technology thus suggest an increasing polarization between the very much higher skills required by system designers, product designers and management, and the degrading and deskilling of the bulk of traditional skills. As Braverman puts it:

> Since, with the development of technology and the application to it of the fundamental sciences, the labour processes of society have come to embody a greater amount of scientific knowledge, clearly the 'average' scientific, technical and in that sense 'skill' content of these labour processes is much greater now than in the past. But this is nothing but a tautology. The question is precisely whether the scientific and 'educated' content of labour tends towards *averaging*, or, on the contrary, *polarization*. If the latter is the case, to then say that the 'average' skill has been raised is to adopt the logic of the statistician who, with one foot in the fire and the other in ice water, will tell you that 'on the average', he is perfectly comfortable. The mass of workers gain nothing

from the fact that the decline in their commond over the labour process is more than compensated for by the increasing command on the part of managers and engineers. On the contrary, not only does their skill fall in an absolute sense . . ., but it falls even more in a *relative* sense (Braverman, op cit, p425).

In our view, the evidence cited above, particularly that of Noble and Wilkinson, shows quite clearly that technology, and the social relations it involves, is flexible and that its final form is a result of struggle, often between different groups of labour and capital, as well as between capital and labour themselves. But would a labour process controlled by the workers be as 'productive' (even if we distinguish between 'real efficiency' and 'capitalist efficiency') as that controlled by capital? 'Capitalist' efficiency may be distinct from other forms of efficiency in a number of ways. For example, the deskilling process (as explained by Babbage), reduces the *cost* of labour, without affecting the extent of labour 'inputs'; in a situation of unemployment, labour-saving innovations may save the capitalist money, but not society's resources since there is no loss of production arising from the employment of workers. Marglin (1976) makes a similar point in discussing the evolution of the factory:

[the division of labour] had little or nothing to do with the technological superiority of large scale machinery. The key to the success of the factory, as well as its inspiration, was the substitution of capitalists' for workers' control of the production process; discipline and supervision could and did reduce costs *without* being technologically superior (p29).)

This, in our view, remains largely untested for the new automation technologies. For this reason the work currently occurring in various engineering workshops around the world to redirect technology, to take advantage of the flexibility offered by electronics and to build on rather than displace the skills of the operator, is particularly important. One of these attempts is occurring at the University of Manchester Institute of Science and Technology (UMIST) where the primary objective is:

To develop software which will enable the operator to program a flexible manufacturing system by making the first of a batch of parts. In doing so, to develop a methodology for the simultaneous consideration of social and technical aspects during the development of new technology (Rosenbrock, 1979).

The UMIST team distinguishes between two paths of development

The first way is the one that is almost universally followed at present. Human subordination is sought in a number of ways. Work is paced by the machine. It is subdivided and rigidly specified so that it takes on a machine-like character, complementary to the operation of the machine. Situations requiring the human qualities of skill and judgement are eliminated by standardisation wherever possible, and where this cannot be done they are removed from the place of work and the necessary decisions taken elsewhere.

Examples of the second way hardly exist at present, though they existed at the beginning of the Industrial Revolution (Hargreave's spinning jenny, Crompton's mule [which 'stemmed from the research workers' implicit assumption that he was designing a machine for himself to use' (passim)]). The human qualities of skill and judgement are not eliminated, but are assisted and made more productive. They are not removed from the place of work. It is accepted that total standardisation is not achievable (machines, for example, may be out of action, or parts required may not be available in the specified order). Human intervention to deal with these abnormalities is assisted by any means which will make it more effective (for example, computers to assist in the rescheduling of work) (passim).

The UMIST team are aiming to utilize the record-playback principle (which we discussed earlier in relation to Noble's evidence) by building on the skills of the operator and his/her interaction with the work process via computers. By so doing they aim to counteract four beliefs widely held in relation to the problems of design of technology, namely technological dterminism (ie there is no alternative route), economic determinism (ie such redirected technologies are uncompetitive), political determinism (ie redirection is impossible without the destruction of the power of capital) and the inherently authoritarian attitude of engineers. Whether the results of this attempt prove to be economically viable, and whether it is possible to redirect technology without major changes in the structure of power, are the two key questions which these experiments hope to uncover. But until this is done, by the UMIST or any other team, it is best to remain agnostic in answering the question of whether labour processes controlled by the workforce are as 'productive' (in their uses of scarce resources) as those controlled by capital.

AUTOMATION AND EMPLOYMENT

The effect of automation on employment is perhaps the most obvious and widely recognized issue when discussing the impact of new technology. Yet strangely enough concern was at its height in the late 1950s and early 1960s and thereafter interest dwindled until very recently. In part, as we argued in earlier chapters, this was because despite widespread forecasts (eg Einzig, 1957) the first generation of computers were associated with an increase in employment rather than the displacement of labour. In addition, the 1950–72 period was a time of long wave upswing, with relatively full employment in most developed economies (see Chapter 1). But now things are different. Most economies are moving into depression and the unemployment rate grew right through the 1970s (Chapter 1). Computing power is beginning to diffuse widely through the manufacturing and service sectors, and unlike the first generation of mainframe computers the second and third generation of mini- and microcomputers are beginning to take a substantial toll of employment in established industries. Not surprisingly, therefore, there is renewed interest in the relationship between automation and employment.

This is an extremely complex problem which defies simple and determinate answers of the 'automation-will-create-an-unemployed-labour-force-of-5 million-in-the-UK-alone' variety to be found in texts such as Jenkins and Sherman (1979). There are a variety of difficulties standing in the way of these projections. The first is that of uncertainty. At this early stage it is not always clear to what extent automation technology will displace labour in particular sectors. Second, granted that automation of existing processes displaces labour, what is the potential provided by new products and processes for creating additional work? Third is the 'income effect' – automation, by displacing labour, cuts costs and hence cheapens commodities: but will it lead to an increase in demand, thereby allowing for almost full employment levels to be maintained? Fourth, there is the trade effect in which the incorporation of electronics into product and/or process may increase global sales and thereby stimulate increased production. And, finally, there is the question of whether there will be a change in consumption behaviour, away from manufactured products (albeit produced in automated factories) to services which are said to be inherently labour-intensive (the so-called 'post-industrial' society).

These five points suggest caution in attempting to assess the impact of automation on overall employment. Yet to veer in the other direction and argue that no attempt can, or should, be made to discuss these crucial issues, is perhaps even more misguided. So with these reservations in mind we shall attempt to plough a path through a troubled and divided debate to see if any broad conclusions can be realized.

We begin first with the micro-sectoral impact of electronics-based automation technologies on employment. It is instructive to take a few examples from key sectors to give a flavour of the magnitude of this displacement. In Germany, one of the most successful of the developed economies in the world, whereas output in the whole of manufacturing plus mining sectors grew by 13.5 per cent between 1970 and 1977, aggregate employment fell by 21.3 per cent; in the key office equipment and data-processing sector, output grew by 48.9 per cent over the period, whereas aggregate employment fell by 27.5 per cent (ETUI, 1980). In the Japanese television industry – uniquely successful in world markets in the 1970s – total employment fell from 48,000 to 25,000 between 1972 and 1976, despite an increase in output of 25 per cent in the same period (Jenkins and Sherman, 1979). In the Japanese electronic consumer goods sector as a whole – which dominates many world markets – aggregate employment fell by 31.2 per cent between 1976 and 1979 (International Metalworkers Federation, 1981). At a more detailed level the trends are similar. Thus in the automobile industry – currently the scene of a large employment shake out – robotic welding increased labour productivity by over five times at Volvo, and by over seventeen times in Fiat (Bessant, 1980). In one study on the impact of robots (Miller and Ayres, 1982) it was found that first generation robots alone are likely to displace around one per cent of the *total* US labour force and second generation robots (with some sort of sensing capability) will displace a further three per cent (this of course, ignores other types of automation technologies described in Chapters 3–6).

These examples – directly related to the use of electronics in automation technology – can be matched by similar statistics drawn from many other sectors, especially those (such as telecommunications) where electronic controls substituting directly for pre-electronic control systems lie at the heart of a process. But the question remains as to what overall significance this labour displacement has, thereby drawing us to a more macro-perspective. In moving to this wider scenario it is essential to see the significance of two major trends. The first is that even prior to the downstream diffusion of electronics into automating manufacturing processes – which as we saw in earlier chapters, was largely a phenomenon beginning in the late 1970s – there had been a tendency for the employment in manufacturing to decline in all of the major economies (Soete 1979). Thus of all the major OECD economies, only the USA employed a greater number in manufacturing in 1979 than in 1970; and even in the USA the rise in the working population over the same period meant that the share of the population employed by manufacturing fell. This phenomenon, as we saw in Chapter 1, was explained by the emergence of economic crisis over the 1970s and in fact partly explains the drive towards labour-displacing automation.

The second overall trend to be aware of is the change in the sectoral nature of employment. As can be seen from Figure 8.1 the share of agriculture in total employment fell in all the major economies between 1965 and 1975. Only in Japan did the share of manufacturing increase, but this was between 1955 and 1971 and thereafter even in Japan this share fell. Thus it was that the service sector was a major source of employment growth in the post-war period.

The extent to which automation technologies will displace labour from the manufacturing sector in the 1980s is difficult to quantify. The critical observation which these attempts to quantify future unemployment rest on, is that even in manufacturing a major (if not *the* major) share of employment occurs in information-processing activities, that is in what we have called the coordination and design spheres of production. For example, in British Leyland, over 60 per cent of the labour force was employed off the shop-floor; evidence from other enterprises suggests similar patterns. The argument is that as intra-sphere automation in the coordination sphere proceeds (Chapter 5), it is precisely these information-processing activities which are most prone to labour displacement. Siemens, the giant German electronics and electrical machinery firm estimate that over 40 per cent of all clerical jobs (including those in other enterprises) would be lost by 1985 due to intra-sphere automation. Despite the possibility that new jobs and skills will be created as electronics-related automation technologies diffuse and as new products are introduced (a subject we will return to), it is difficult to avoid the conclusion that the 1980s are likely to see the net displacement of workers from the manufacturing sector. Most observers concur, even those who take a more optimistic view of overall employment prospects. The real issue is how great will this net displacement be? On the evidence of Chapters 3–6, even when set against the prospects of new job creation, we think that the net displacement effect is likely to be substantial.

The scope for job displacement from the agricultural sector varies. Some economies such as the UK and US have such a small proportion of the labour force employed in this sector (currently less than five per cent of the labour force) that however great the proportionate loss of employment might be, the

Fig. 8.1 Sectoral employment in seven economies

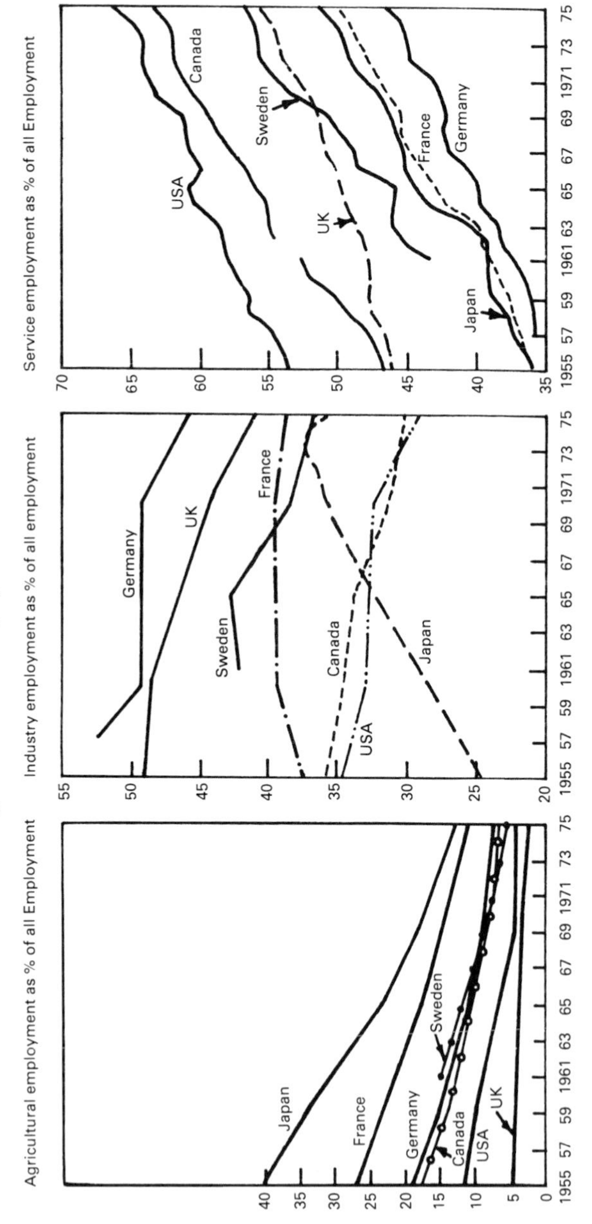

Source: Gershuny, 1979

overall impact of this displacement will be small. However, in other countries the picture is somewhat different and in Austria, Ireland, Italy, Greece, Spain and Portugal between 12 and 30 per cent of the labour force is still employed in agriculture. The pressures of greater competition (as the entry of some of these economies into the EEC takes effect) together with the emergence of new automated technologies in agriculture, is likely at the very least to stop the agricultural sector from absorbing a larger number of workers.

It is the service sector which is believed by some to offer the potential for mopping up the labour displaced from industry and agriculture, for as Bell (1974) argues, in the same way that industry took up labour displaced from agriculture in the nineteenth century, so it is possible that the service sector will compensate for labour displaced from industry in the latter half of the twentieth century. It is here that the evidence has begun to shake even those observers. There are two factors at work which tend to undermine their optimism. The first is that a disproportionate number of service sector jobs involve information-processing activities and it is these, as we have seen, that are likely to be most susceptible to displacement by automation technologies. A case in point is the banking and insurance sectors where information-processing automation technologies are already beginning to have an impact, as is shown in Table 8.1 below, but where the full potential impact of automation technologies such as electronic funds transfer (EFT) have not yet been realized. Other potential areas are health (heavily dependent upon information processing) and retailing where electronic point-of-sales cash registers linked to inventory control systems and automatic warehousing offer substantial productivity gains. It is worth re-emphasizing that the discussion in earlier chapters has deliberately confined itself to industrial activities. However most of the discussion applies equally to the service sector (see Chapter 2).

Table 8.1 Employment in banking and insurance in 5 EEC countries

	Thousands			*percentage annual change*	
	1964	*1974*	*1977*	*1964–74*	*1974–77*
Belgium	72	190	194	10.2	0.7
Denmark	78	133	136	6.1	0.7
Fed. Rep. of Germany	790	1131	1065	3.7	−1.9
France	577	1060	1161	6.3	3.1
United Kingdom	944	1306	1328	3.3	0.6

Source: ETUI 1981

The second factor undermining the ability of the service sector to mop up labour displaced from other sectors is the emergence of what Gershuny (1979a) calls the self-service economy. By this he means the tendency for consumers to buy 'domestic capital goods' instead of buying in these services from outside the household:

> . . . we cannot expect service employment to grow naturally and of its own accord. The sizeable areas of past growth in service consumption in the developed world have been in education and medicine. Other service consumption – transport, entertainment, domestic – has fallen considerably over recent decades. Rather than buy increasingly extensive services, households have preferred to buy even cheaper goods which are used to produce approximately equivalent results – cars instead of transport services, domestic machines instead of domestic services, television instead of cinema (Gershuny, 1979a, p1).

Figure 8.2 provides strong evidence for this latter point.

At an aggregate level, therefore, we see a tendency for labour to be displaced from each of these three major sectors. Indeed much of this preceded the widespread diffusion of electronics-related automation technologies (reflecting other 'non-technological factors' discussed in Chapter 1) which are liable only to exacerbate the problem. To compound the issue the 1980–5 period is likely to see disproportionately high growth in the European labour force of around 0.9 per cent pa compared to 0.6 per cent in 1974, 0.5 per cent in 1975, 0.3 per cent in 1976, 0.4 per cent in 1977, 0.5 per cent in 1978, 0.6 per cent

Fig. 8.2 Substitution of goods for services in consumption

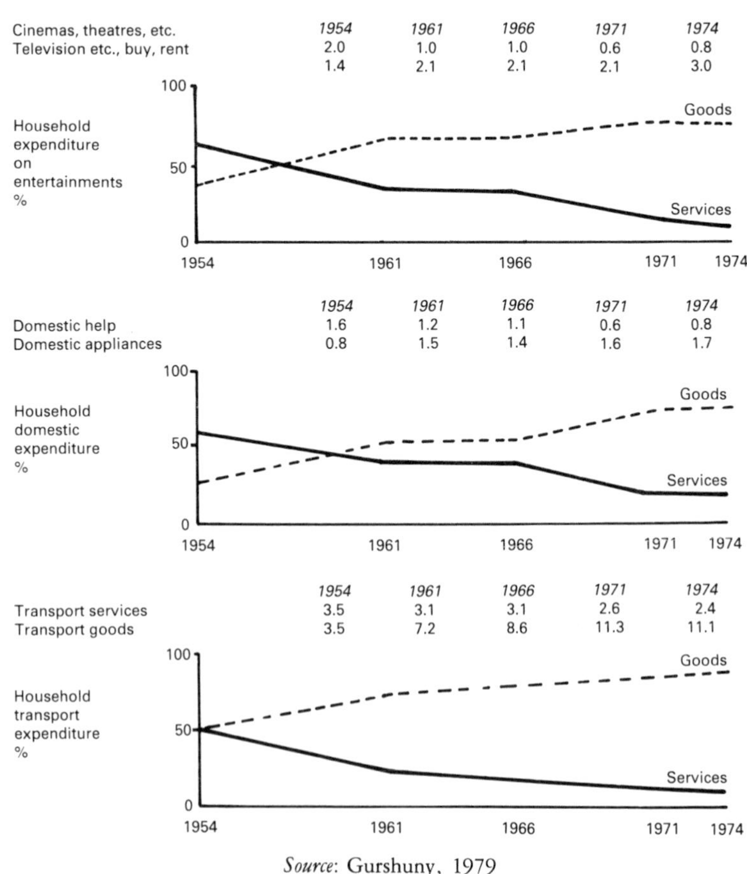

Source: Gurshuny, 1979

in 1979 and 0.7 per cent in 1980. This is because of a slowdown in the number of men reaching retirement age due to the low birth rate during the 1914–18 war (ETUI, 1981).

So what does this portend for the overall level of unemployment? The first thing to note is the importance of historical perspective. Fullish employment as we know it is a fairly recent phenomenom and, if anything, is historically atypical. For example, in only one year between 1919 and 1939 was the British rate of unemployment less than 10 per cent (in 1927 it fell to 9.6 per cent): over the whole time period it averaged 14.2 per cent (Beveridge, 1944); in the US, in the 1930–40 period, the number of unemployed fell below 8 million only once – even in 1938 20 per cent of the labour force was unemployed (Galbraith, 1959). In this context it is interesting to quote the views of the 'father of cybernetics', Norbert Wiener, on the likely impact of automation on employment:

> It is perfectly clear that this will produce an unemployment situation in comparision with which the present recession and even the depression of the thirties will seem a pleasant joke (Wiener, 1950, p189).

This quote from Wiener dates back to the late 1940s. However, a number of more recent attempts have been made to project the likely rate of employment in the 1980s. A minority are optimistic: the 'Think Tank' of the UK government, for example, believed that market growth will balance out the jobs lost through automation (CPRS, 1978). But most observers are less sanguine. The problem inevitably lies in giving credence to the figures which are produced. At one extreme are the 'back of the envelope' calculations of Barron and Curnow (1979). Using 1970 US data on occupational structure assessing that 'high risk jobs' (their classification) face a 50 per cent chance of displacement, 'medium risk' have a 25 per cent chance and 'low risk' jobs a 10 per cent chance, they compute an incremental unemployment rate for the USA (that is over and above the existing 1979 rate of around 6 per cent) of 18.2 per cent. A similar exercise for the UK yielded an incremental rate (in addition to the 1979 rate of 5.6 per cent) of 16 per cent.

Such a procedure is inevitably highly speculative. Yet even more cautious attempts at prediction yield only slightly less disturbing figures. One attempt by the Institute of Manpower Studies at the University of Sussex made the cautious assumption that the output/employment ratio in the UK would be the same in the 1980s as it was in the early and mid 1970s (that is there would be no speed up in the rate of technical change). However as intra-sphere and inter-sphere automation proceeds, the growth of labour productivity (especially in the coordination sphere of production) is likely to grow rapidly in the 1980s. On this basis, therefore, the no-change-in-productivity-growth assumption of the IMS team is disconcertingly conservative, and any increase in this rate is only likely to raise the unemployment rate further. The IMS team also assumed an almost certainly optimistic growth rate of 2.75 per cent between 1979 and 1991, concluding that the unemployment rate in 1991 would be around 12 per cent (Jenkins and Sherman, 1979).

The truth is difficult to determine, not least because the end result is not pre-ordained, but will be a consequence of political struggle (between labour and capital) and economic struggle (between different capitals). Unless there is a very large and unexpected change in the world economy towards higher rates of growth (which, as we observed in Chapter 1, looks unlikely) we can anticipate rates of unemployment in excess of the 12 per cent estimate of the IMS which, as we saw, was based on the unrealistic assumption that there would be no speed-up in the rate of technical change in the 1980s. In the light of this rather disturbing conclusion it is worth examining briefly three factors which might act to counter this drift to substantially higher levels of unemployment: new products and skills, a shorter working week and opposition by labour.

New products and skills

By their nature it is difficult to predict the emergence of new products since if they were predictable many of them would already be in production. However, it is possible to make some observations. First, by hypothesis, we argued in Chapter 1 that whereas the 1950–72 period was one of new-product-led long wave upswing, the post-1973 period saw the transition to a phase of rationalizing downswing in which the heartland technology, microelectronics, would be utilized primarily for cost reduction rather than product innovation. We believe that the evidence cited in Chapters 3–6 supports this hypothesis. Second, the evidence of new product development in the 1970s would suggest that these new products, particularly when they are allied to electronics technologies, do not on balance create employment, especially when they substitute in part for more labour-intensive technologies (eg video tape recorders for the cinema). Recall the example of the Japanese electronics industry which saw not only growing global market shares, but also new products (such as video tape recorders) over the 1970s, yet registered a 31.2 per cent decline in aggregate employment in the 1976–9 period alone. And third, there is the question of a countervailing increase in employment as a consequence of introducing new automation technologies. Indeed Rothwell (1981) offers this as one possible explanation for the exceptional performances of the US economy in the 1970s in which employment in the manufacturing sector was less badly affected than in other industrialized economies. However, an equally plausible explanation might be that the less-bad US employment performance was more likely due to the failure of their older, pre-electronic firms to introduce new automation technologies, and to the less substantial and delayed impact of the post-1973 recession. Moreover, as Rothwell points out, the evidence from Western Europe and Japan where new technology tended to be introduced by both the older and the large firms, does not confirm this hypothesis.

And finally, there is the possibility that the introduction of new technologies, whilst displacing direct labour, creates ancillary jobs. The evidence on this is not very convincing however. In one detailed study conducted by the British Civil Service, it was estimated that, whereas with existing technology staff would be increased from 2,461 to 3,393 over the next decade, under a computerized system the increase would be kept to only 31 people. In the same period the number of data-processing staff would rise from 21 to 28 under a manual system and from 21 to 172 under a computerized system (Basset, 1979). And earlier in the chapter we reviewed the evidence of Bright and others which reached a similar conclusion with respect to the number of support jobs which were likely to be created in the manufacture sphere of production.

Shorter working week

An obvious and logical conclusion to the above discussion is a reduction in the working week to enable more people to be employed and in such a way that more time would be available for leisure. Indeed it is often forgotten that this is supposed to be a major goal of economic development and should consequently be a preferred solution. However, there are problems with this strategy. First, the unemployment rate is not terribly sensitive to a reduction in working hours. Jenkins and Sherman conclude that:

> Various studies have been and are being carried out to quantify the impact on work of such a cut, but in terms of the new technologies, it cannot be more than a one to two relationship at best. In other words, for every 2 per cent reduction in hours only 1 per cent more jobs will be saved or created (Jenkins and Sherman, op cit, p154).

But more importantly unless the reduction in working week was compensated for either by a reduction in real wages and/or a matching cut in the hours worked in competing countries, severe problems would arise with respect to international competitiveness. (We return to this in Chapter 10.) Thus a shorter working week, whilst undoubtedly offering potential, can only be seen realistically in terms of matching policies with respect to real incomes and trade negotiations. It is thus not a policy which can be pursued adequately at the national level alone.

Unemployment as a consequence of struggle

The advance of crisis over the past decade was associated with a growth in conflict between labour and capital and between different capitals. It is this which largely explains the drive towards automation by capital (Chapter 1). However, as critics of Braverman have pointed out, it is not wise to merely accept that capital will always win. Indeed there are many examples, particularly in the UK, in which, through a process of extended struggle, the trade unions have managed to prevent or slow down the introduction of new labour-displacing automation technologies. The problem with this, as we have seen in previous discussion, is that it leads to a loss of international competitiveness and hence an eventual loss of work (and output) anyway. For as Jenkins and Sherman, arguing from a perspective of trade union response, have put it,

> *Remain as we are, reject the new technologies and we face unemployment of up to 5.5 million by the end of the century. Embrace the new automation technologies, accept the challenge and we*

end up with unemployment of about 5 million . . . What is clear is that whichever road we take work will collapse [their emphasis] (Jenkins and Sherman, op cit, p113).

They might also have added that the first alternative is also associated with a substantial loss in output and hence a decline in real standards of living.

In summary, therefore, we are drawn to the conclusion that the introduction of new automation technologies, associated as they are with the deepening of economic crisis, is likely to lead to high and sustained levels of unemployment, probably in excess of 12 per cent of the labour force. The countervailing tendencies offered by new products, the demand for new skills, the introduction of a shorter week and the resistance to new automation technologies all hold little prospect for substantially altering this perspective.

Before we turn to a discussion of the alternative prospects available (Chapter 10), we will continue our analysis of the likely impact of automation technologies by assessing the implications for the Third World (Chapter 9). For not only is this important in the static welfare sense, but also it is relevant to the restructuring of production at the global level and the implications this has for the diffusion of the new automation technologies. However, prior to this, it is desirable to conclude the discussion in this chapter on the impact of labour, by disaggregating 'labour' into three important sub-divisions, namely gender, age and region.

With respect to *gender* it is a characteristic of most industrialized economies that women are disproportionately employed in lower-paid information-processing (eg secretaries) and service sector (eg health workers and waitresses) jobs. These, as we have seen, are likely to be most affected by the introduction of automation technologies in the coordination sphere of production and into the service sector. An additional overlay is that the patriarchal system which characterizes these economies, means that women's employment is considered more marginal. They are said to work for 'pin-money' as opposed to the male 'bread-winners' of the family. The consequence is that the impact of automation on female employment is likely to be disproportionately severe. In fact, we have already seen these trends emerging:

In 1973 the women's unemployment rate in EEC countries was 2.3 per cent compared with the male unemployment rate of 2.5 per cent; at that time women made up 33.1 per cent of total unemployment. In January 1980 the unemployment rate amongst women was 7.2 per cent compared with 5.3 per cent for men; at that time the proportion of women in total unemployment had risen to 44 per cent (ETUI, 1981, p7).

Youth are similarly likely to be disproportionately affected by the introduction of new automation technologies since the introduction of these technologies is often associated with a reluctance to take in new labour, even when they do not involve redundancies of existing employees. Again, as is shown in Table 8.2 below, the evidence already exists to support this conclusion.

And finally, it is important to take note of the changing *regional distribution* of unemployment. In the UK, the older labour-intensive industries such as the car and machine tool sectors have been particularly badly it, either directly

Table 8.2 Youth unemployment rates (%)

	1973	*1979*
France	6.3	13.3
Germany	1	3.8
Italy	11.9	24.6
UK	2.8	11.9
All EEC Youth as a proportion of all unemployed	28	36.4[a]

([a]) 1978
Source: Calculated from ETUI, 1981.

by the introduction of new automation technologies, or indirectly by their introduction in competing countries. And the automation technologies which have been introduced in the newer electronics-based sectors have tended to be sited in the south west, south east and parts of Scotland. The result has been a 'collapse of work' in the traditional industrial heartlands as can be seen from Figure 8.3. Similar trends have emerged in almost all of the industrialized world including the USA (Miller and Ayres, 1982).

Fig. 8.3 The pattern of regional unemployment in the UK, April 1983

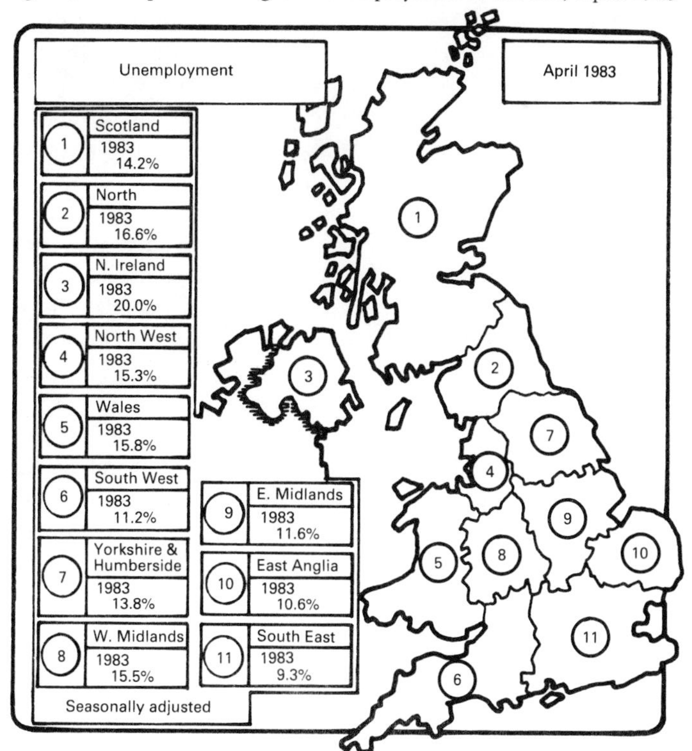

Source: *Financial Times*

CHAPTER NINE

Impact on the Third World

THE THIRD WORLD IN THE POST-WAR ECONOMY

The discussion so far has largely been focused on the emergence of automation technologies in a group of highly industrialized and rich developed economies. Yet the twenty major 'market' economies – that is, the USA, Japan and Western Europe – which have been and are likely to be the major actors in this unfolding story, between them account for only around 16 per cent of the world's population. They nevertheless produce around two-thirds of world GNP and account for around 65 per cent of global industrial value added.

In Table 9.1 we observe some of the major differences between various sets of the world's economies, subdivided according to the World Bank Development Report (1982): low-income countries (including India and China) whose per capita incomes were lower than $420 in 1980; middle-income countries (per capita income between $420 and $4,600); high-income oil exporters ($8,640 to $27,000); industrial ($4,880 to $16,440); and nonmarket economies (ie Eastern Europe, whose estimated per capita incomes range between $3,900 and $7,180). Aside from the imbalance in population, incomes and industrial output noted above, it is instructive to observe the disturbing disparities in average levels of food consumption, life expectancy, child mortality and access to safe water.

In recent years, many developing economies have come to experience growing shortages in the domestic production of food required to feed their populations. Excluding China and India, the index of food production per capita in low-income economies actually fell by five per cent during the periods 1969–71 and 1978–80. There is consequently an urgent need for many Third World countries to expand their agricultural sectors. Moreover, when relating these 'developmental imbalances' to the structure of these economies, a clear conclusion emerges. That is, unless a country is 'lucky' enough to have abundant supplies of oil (arguably of little benefit to the mass of the population), then wealth is associated with a well-'developed' industrial structure (and especially the manufacturing subsector as distinct from mineral industries and utilities) and a skill-intensive set of service industries. Thus the role to be

Table 9.1 Some indicators of different types of economies (a)

	Population (millions) 1980	*GNP per* capita ($) *1980*	*Adult literacy 1977*	*Life expectancy at birth*	*Per capita energy consump-tion 1981 (d)*	*Infant mortality rate (age 0–4) (e)*	*% with access to safe water 1975*	*Daily calorie supply 1977*	As percentage of GDP (1980)			
									Industry	*Manufactu-ring*	*Agriculture*	*Services*
Low income countries (b)	2,161	260	50	57	421	106	31	2,238	35	15	36	29
Middle income countries	1,139	1,400	60	60	965	91	50	2,561	40	19	15	45
High income oil exporters	14	12,630	25	57	2,609	113	88	NA	77	4	1	22
Industrial market economies	714	10,320	99	74	7,293	12	(f)	3,377	37	27	4	62
Non-market industrial economies (c)	353	4,640	100	71	5,822	26	(f)	3,489	63	NA	15	22

(a) All averages weighted
(b) Includes China and other low income socialist countries.
(c) All E. Europe.
(d) Kilograms of coal equivalent
(c) per 1,000 live births
(f) No information given, but estimated at over 95 per cent.

Source: IBRD (1982), Tables 1, 3, 7, 21, 22.

played by the industrial (and especially the manufacturing) sector in future economic growth should not be discounted. What, then, has this role been so far?

It was only in the two decades after the second World war that serious attempts at industrialization were made in most of the Third World, particularly in those countries which had been colonies of European powers. In general, the path states tended to choose to expand industrial output was that which substituted domestic production for goods which had earlier been imported, so-called import-substituting industrialization (ISI). But by the end of the 1960s much of the gloss had worn off these ISI strategies. Evidence began to accumulate that such industries were 'inefficient', producing high-cost and low-quality output. Moreover, the output and foreign exchange gains arising in the early 'easy' stage of ISI became increasingly difficult to sustain, and most countries pursuing this path of industrialization began to face mounting balance of payments deficits and reduced growth rates. Perhaps as serious was the change in the terms of trade between agriculture and industry – that is, manufactured goods became increasingly expensive compared with agricultural commodities. The consequence in many less developed countries (LDCs) was a decline in the growth in agricultural productivity and a greater inequality in incomes. This occurred partly because agricultural inputs (which were produced in the industrial sector) became more expensive, and partly because declining terms of trade reduced incentives to farmers.

However, by this stage – the early 1970s – there was growing awareness of the gains obtained by some economies from a different industrial strategy, that which concentrated on the export of manufactures. Both their overall growth rate and their growth of manufactured exports were extremely high, and because of this, the share of the Third World in global manufacturing value-added began to expand. In 1965 it had been around 7 per cent, rising to nearly 10 per cent over the next decade. Perhaps more significantly, most of these manufactured exports appeared to be going to the industrialized economies (Table 9.2) and the earlier emphasis on traditional manufactures such as shoes, leather, garments and textiles was undermined by the expansion of technology-intensive products. Particularly important was the progress of electrical and electronic products. The share of these products in total LDC manufactured exports to developed countries rose from 4.8 per cent in 1967 to 15.4 per cent in 1974 (Plesch, 1978). However, despite these changes, over 53 per cent of total LDC

Table 9.2 Percentage of imports of manufactures coming from developing countries

	1962	*1970*	*1975*	*1978*
All industrial countries	5.3	6.8	10	13.1
Europe	4.2	4.8	7.5	9.6
Germany	4.6	6.3	10.8	12.9
Japan	5.9	11.4	21.4	25.1
USA	12.3	14.7	21	26.7

Source: IBRD, 1982.

exports to developed countries in the late 1970s remained 'traditional products', that is, shoes, leather products, garments, textiles and plywood.

It is important to bear in mind the uneven nature of this quite remarkable expansion of Third World manufactured exports. The 'success' was limited to only a narrow range of developing economies, usually called the newly industralized countries (NICs) of which the largest eight accounted for 77 per cent of total manufactured exports in 1976. Of these, four stand out in importance, namely, South Korea, Taiwan, Hong Kong and Singapore. For example, in the electrical machinery and electronics subsectors relevant to our interest in automation, the four largest exporters (Singapore, Taiwan, South Korea and Hong Kong) actually increased their share from 60.7 per cent in 1967 to 76.9 per cent in 1978 (Clark and Cable 1982). At the same time, much of the export expansion was attributable to the locational decision of transnational corporations (TNCs). As we observed in Chapter 1, these firms responded to the emergence of economic crisis at the centre of the world economy by siting parts of their productions process in developing economies where wages were low, labour was compliant, government incentives were generous (generally offering ten-year tax holidays) and two-and three-shift capacity utilization was feasible. When transnational corporations were not involved in actually producing in the Third World, their presence was increasingly felt as purchasing agents, effectively covering the marketing of much of LDC manufactured exports (Hone, 1974). And when they were involved in production, in a selected number of 'export platform economies', they were disproportionately represented in exports: in South Korea the share of transnational corporations in production in the mid 1970s was 11 per cent, but they accounted for 28 per cent of exports; in Singapore the respective ratios were 30 per cent and 70 per cent. The consequence of all this was a substantial increase in 'related party trade' (that is, trade between firms in which there is at least a 5 per cent link in equity holdings between the parties). In 1976 this accounted for the following shares of US imports of electrical machinery and electronic products from major LDC exporters: Malaysia (98 per cent), Mexico (97 per cent), Brazil (85 per cent), Singapore (92 per cent), Taiwan (68 per cent), Korea (67 per cent) and Hong Kong (45 per cent).

Leaving aside the unevenness of this export performance over the 1970s, it is instructive to observe its impact on industrial strategies in the coming years. Flushed with the success of the newly industrialized countries, most LDCs and multilateral aid agencies are aiming to continue, or to emulate, this outward-oriented industrialization in the 1980s. For example, between 1978 and 1980 the number of Free Trade Zones increased from about 220 to over 350, most of which were in the Third World including even such unlikely countries as China. One of the forums for developing countries, the United Nations Industrial Development Organization (UNIDO), convened a gathering of governments in 1975 to issue the Lima Declaration, in which a goal was set for the Third World's share of industrial value added (a shade under 10 per cent in 1976) to expand to 25 per cent by the year 2000. The World Bank and the International Monetary Fund also jumped on the bandwagon, making aid conditional upon an outward-oriented industrial strategy. For example, in its policy document for Africa in the 1980s, *Accelerated Development in Sub-Saharan Africa: an Agenda for Action* (IBRD, 1981), the World Bank offers two major strategies: the stimulation of agricultural production and the expansion of exports. Particularly high on the list of potential export items are manufactured products; in the case of Kenya, for example (Godfrey, 1983), the imputed growth rate for manufactured exports is 28 per cent pa over the 1980s compared with a historic growth rate between 1973 and 1980 of only 6.8 per cent pa.

Thus, largely because of the undoubted success of a limited number of newly industrializing countries in the export of manufactures in the 1970s, a key role is being given to the manufacturing sector in the development strategies of the 1980s, particularly in relation to the expansion of manufactured exports. But, as we have seen in earlier chapters, it is precisely in this sector in the industrialized economies that the new automation technologies are diffusing. This potential conflict brings us to the substance of this chapter, namely, to explore the likely impact on LDCs of the diffusion of new microelectronics-based automation technology. We begin the analysis by examining the impact on outward-oriented industrialization strategies, and then proceed to a brief examination of the use to which the new technology is being put in meeting basic needs in the Third World.

THE IMPACT OF AUTOMATION ON STRATEGIES FOR EXPORT-LED GROWTH IN MANUFACTURES

Inevitably, the potential for export-led growth strategies (in fact we are referring to strategies involving the export of manufactures but use this phrase as a convenient shortening) is affected by a variety of factors, not all of which concern the utilization of automation technologies. While our attention centres on these technology-related factors, it is valuable first to consider briefly other potential influences on the viability of these growth strategies. Four stand out in importance.

First is the problem of the growth of the world economy, a factor we considered in Chapter 1. After 25 years of sustained post-war growth the global economy appeared to move to a significantly lower growth path after the mid-1970s. The consequence was a reduction in the expansion of world trade, such that the 1970–8 average growth rate of 6 per cent pa declined to 5.4 per cent in 1978, 5.9 per cent in 1979, 1.5 per cent in 1980, and zero in 1981. Indeed, the volume of trade actually fell by 2.1 per cent in 1982. Prospects for growth in the world economy in the 1980s seem bleak, making it likely that, even with no other intervening factors, the growth of LDC manufactured exports will decelerate.

Second is the problem of LDC debt. This grew from $383 billion in 1979 to $598 billion in 1982. On an annual basis, the aggregate debt-service burden (that is the rate of annual debt plus interest payments to total exports) for the oil-importing LDCs was over 20 per cent in 1982: in the case of Mexico and Brazil the ratio in 1980 was 34 and 32 per cent respectively. This increasingly severe debt problem can have two potentially conflicting influences on LDC manufactured exports. Either it can enhance them as countries try to increase their exports to cover their balance of payments deficits; or the enormity of the foreign exchange problems can induce restrictive policies on imports and general economic activity which effectively undermine the ability of the economy to export successfully. Recent events in Mexico suggest that this latter effect is a real danger for many LDCs. Moreover, the type of aid now being offered by the International Monetary Fund (IMF) and the World Bank, which involves a reduction in protection and a diminution of state-support for industry, is likely to have the effect of hampering the growth of manufacturing production, leading to a tendency towards 'de-industrialization' in many LDCs.

Third is a problem which has come to be called the 'fallacy of composition'. That is, whilst an export-oriented strategy may be viable for a small number of LDCs, it cannot conceivably be viable for *all* or for *many* LDCs. It may, for example, have been possible for Korea's electronics exports to grow at an annual compound growth rate of 37.8 per cent between 1971 and 1981, but it was patently impossible for many other LDCs to show the same growth rate in the electronics sector. The only circumstances in which this might have been possible at an aggregate level would have been if there had been a technological-ladder effect in which individual economies increasingly moved to higher technology sectors as LDC economies entered production. But the existence of unemployment in the advanced economies has made it increasingly unlikely that they will 'vacate' mature sectors to provide space for newcomers. The resurgence of production in traditional sectors such as garments (Hoffman, 1983), steel and textiles, illustrate this well. In France in 1981 alone, for example, the state provided subsidies for the textile sector of around £200m.

Finally, LDCs face an increasingly militarized world. The geo-political rivalry between East and West is foreclosing many industrial options which would have made it easier for LDCs to develop an appropriate industrial base and to expand their exports of manufactured products in the 1980s. Nowhere is this better illustrated than in the Indian Ocean, where the changes in US submarine-based-missile technology have moved the arena of conflict from the Arabian Gulf to the equatorial zone (Kaplinsky, 1983b). This is having major consequences for all the surrounding economies, and in particular for India. Faced with the delivery of sophisticated F-16 planes to a potentially nuclear-powered adversary, Indian industrial strategy is being significantly affected, such that increasing dependence is being placed upon foreign technology and 'inappropriate' industries.

Yet, despite the importance of the above-mentioned factors, each of which has a potentially significant influence on LDCs' attempts to expand their exports of manufactures in the 1980s, it is the technology-related factors which draw our interest. Here we concentrate on three major items: recession and market entry; 'comparative advantage reversal'; and the emergence of different families of technology. We treat each in turn.

Market entry

Post-war economic growth, especially in the industrialized economies, was accompanied by a more than proportionate growth in trade, especially of manufactured products, in which exports grew twice as fast as production. Hence in the 1953–1977 period, the value of global trade in manufactures grew 673 per cent, compared with 513 per cent for trade in all commodities, and only 317 per cent for the output of all manufactures. Inevitably the 'rules' under which trade occurred and grew had to be changed to reflect its growing importance. Of primary importance was the generalized reduction in tariff rates, a process administered in seven rounds by the General Agreements on Trade and Tariffs (GATT). Initially, tariff reductions were conducted on a country-to-country basis and then extended, on the principle of non-discrimination, to other countries. Then with the two most recent rounds (the 1964–7 Kennedy Round and the 1973–9 Tokyo Round) the process became generalized.

An important component in these various rounds of tariff negotiations was the explicit recognition of the need to offer special assistance to LDCs. In the Tokyo Round, for example, a commitment was made to introducing:

> differential measures to developing countries in ways which will provide special and more favourable treatment for them in areas of the negotiation where this was feasible and more appropriate (Quoted in Commonwealth Secretariat, 1982, p29).

As the 1970s wore on, and unemployment and balance of payments deficits became endemic in the advanced countries (see Chapter 1), so the pressures towards protecting industries became increasingly irresistible. (Indeed, as we have seen, in the 1981–2 period aggregate trade actually fell despite an increase in global output.) But, since there was a general commitment to reducing tariff barriers, some other form of entry barrier was required, the result being a plethora of 'non-tariff barriers'. These included 'voluntary export restrictions' (VERs), 'orderly marketing arrangements' (OMAs), quotas, preferential government purchasing policies, local content requirements and subsidies. In fact, GATT has recorded over 600 different types of non-tariff measures developed to stem inroads into domestic markets made by foreign producers.

Because of their relative political weakness, their inability to exchange significant trade concessions, and their concentration in labour-intensive sectors, LDCs become particularly vulnerable to these protectionist pressures. So, despite the good intentions enunciated in the Tokyo Round, the resultant combination of tariff and non-tariff barriers began to hit the LDCs particularly hard as the decade progressed. As the Commonwealth Secretariat Report by a Group of Experts concludes in relation to tariff barriers:

> Tariff cuts were proportionately greater on products traded among the nations of the OECD than they were on those originating in developing economies. In particular, tariffs in the developed countries remained relatively high in comparatively low skill labour-intensive products . . . (Commonwealth Secretariat, 1982, p23).

And in relation to non-tariff barriers, the Group of Experts concluded, 'Where discretion replaces rules, the weakest invariably lose most.' Moreover, the Report continues:

> It is ironic and tragic that a new protectionism should be appearing in the developed countries at a time when there has been a trend towards increasing trade liberalisation and outward orientation in the developing ones, particularly in the newly industrializing countries (op cit, pp32–3).

Thus, the formulation of export-led growth strategies in the LDCs reflects a world in which trade barriers were being removed and global transactions were expanding rapidly. As these conditions change, so a basic premise of the outward-oriented strategies – namely market entry – is being undermined and export-led strategies are increasingly vulnerable. But it is important to bear in mind the link between market entry and technology, since to a large extent the increasing obstacles to trade reflect the emergence of electronics-based automation technologies. This occurs in a number of ways. First, post-war economic expansion was associated with the introduction of new products based upon the new technology. As the long wave proceeds to its downswing, so the heartland technology is diffusing with rationalizing investments and is associated with a descent into economic crisis (Chapter 1). Second, it is a characteristic of the new technology that it displaces labour, further stimulating the public outcry in developed countries for protection from foreign producers. And third, the sectors in which automation technologies seem to be diffusing most slowly (see later) are those which produce 'traditional' labour-intensive products such as garments, textiles and leather products, in which LDCs excel. Hence further penetration by LDC producers in the developed countries markets is likely to have a disproportionately severe impact upon employment patterns in developed economies. All the portents are that as the economic crisis in the advanced countries progresses, and as the new automation technologies diffuse more widely, so LDC exporters face growing difficulties in gaining access to markets in advanced countries.

'Comparative advantage reversal'

Through the 1960s and 1970s, as we have seen, LDCs came to develop a comparative advantage in selected sectors of manufacturing. In addition to the established 'traditional products' such as shoes, leather products, garments, textiles, and plywood, expertise was built up in a limited number of countries in other areas such as electrical engineering and electronics products, basic chemicals and engineering goods. It is often thought that this comparative advantage arose from low wages, but other related factors such as longer working weeks, multiple shift utilization, the suppresion of worker militancy, tax holidays and the flexibility of work practices (given the absence of an industrial heritage) were probably as important. Another common misconception concerns the overwhelming role played by transnational corporations, but as we have seen this varies by country and by industry. In the particularly rapidly growing newly industrialized countries domestically owned industry became increasingly powerful over the 1960s and 1970s, a phenomenon which owed much to the interventionist role played by their governments in limiting the role of direct foreign investment and aiding the expansion of indigenous firms. A third major factor explaining the increasing sophistication of LDC industrial exports was the investment placed in education such that many LDCs now have more than adequate supplies of skilled workers, although there is often a vast mismatch between their skill-requirements and the output of the educational system.

Yet, despite these undoubted competitive advantages, it must be remembered that the success of export-oriented industrialization occurred in the context of an expanding world economy in which the upswing of the long wave saw relatively full employment and a continuous expansion of world trade, particularly in manufactures. The transition from this expansionary upswing to a crisis-laden downswing has, as we have seen in previous chapters, been associated with the diffusion of electronics-based automation technologies, in many cases to sectors in which LDCs either have already, or hoped to develop a comparative advantage. These technologies appear to offer significant benefits to user-firms, particularly in relation to those factors which underlie LDC comparative advantage. For example, they are labour-saving, hence undermining the advantage of cheap labour. Moreover, on balance they use less-skilled labour, hence lowering the average wage cost as well as reducing labour requirements. They can work on a multiple-shift basis, particularly as palletized loading systems are developed (recall the Fujitsu Fanuc factory which operates with 100 labourers on the day shift and only one at night).This saves capital, as well as labour costs. In addition, the technologies enable product lead-times to be substantially reduced with a greater heterogeneity of output achieved without any significant reduction in scale economies. At the same time, product performance can be optimized such as in the contemporary design of fuel-efficient motor cars. And finally, in recapping the major benefits the technology provides to firms which innovate successfully, the control by management over labour is strengthened, hence further reducing production costs in the innovating enterprises.

All of this suggests that if LDC firms do not take up and make successful use of the new automation technologies, then much of their comparative advantage, which seems to have expanded so successfully over the past decade, may be undermined. The key question, therefore, concerns the diffusion of these automation technologies to the Third World. But since the maturation of intra-sphere automation (linking up separate sets of intra-activity automation in each of the three spheres of production) and inter-sphere automation (linking up activities in the different spheres) has been such a recent phenomenon, it is difficult to reach definite conclusions on their future rate of diffusion. We can, therefore, at best only speculate on the emerging patterns of diffusion and their likely impact upon global patterns of trade in manufactures.

So far, there is only limited evidence on *the diffusion of automation technologies*, and that which exists is mixed. Considering the diffusion of CAD equipment (intra-activity and intra-sphere automation in design), sales to LDCs are very limited (Kaplinsky, 1982b). Of over 8,000 turnkey, interactive systems sold, only 32 went to developing countries. Of these some went for terrain-mapping in Central America, backed by US aid and used primarily for counterinsurgency purposes; another large component was their use by transnational oil companies for oil exploration in the Third World. Only in the case of India, where the scientific base is impressive but industrial linkages are very weak, and Brazil, where the automobile industry is expanding, can it be said that CAD technology found receptive customers. However, in 1982 the largest CAD supplier, Computervision, established a sales subsidiary in Singapore. This was with the clear intent of expanding sales in South East Asia, especially in the electronics subsector, which as we saw in Chapter 3 was the major area of expansion of the technology in developed countries in the early 1970s.

A second set of data concerns numerically controlled machine tools. Again we find scant evidence of widespread penetration in LDCs. Jacobsson (1982) has estimated the stock (that is total accumulated sales) of numerically controlled machine tools in four LDCs, which totalled 130 in India in 1979, 325 in Argentina by 1980, 649 in Brazil by 1979 and 1,000 in South Korea by 1980. By comparison, in Sweden (a smaller economy that India, Brazil and South Korea) the stock of numerically controlled machine tools was 4,000 in 1979; in the USA, which, although a much larger economy has been relatively laggard in the introduction of this technology, annual sales in 1968 were 2,917 units, rising to 8,856 in 1980; and in the UK, another laggard in introducing the technology, 1981 sales exceeded 1,550 units. Jacobsson suggests that such comparisons may not be very helpful since they ignore the size of the industrial sectors which make use of numerically controlled machine tools. Relating the stock of these intra-activity automation technologies to the aggregate sales of the nine branches making heavy use of them, he finds the intensity of use greater in South Korea (645 machines per $b) than Sweden (460 per $b). However, without considering the value-added in each country – given that the import – sales ratio is likely to be much higher in South Korea than Sweden – or the complexity of machine tools involved, it is difficult to draw general conclusions from these ratios.

In the case of South Korea, where the electronics industry expanded at around 30 per cent pa between 1973 and 1981, there is little evidence of automation equipment being produced domestically. Of the total 1981 industry output of $3,791m, only 13 per cent comprised industrial products. Within this the major shares were accounted for by wire communication equipment (8.3 per cent), desk-top calculators (1.6 per cent) and radio communication equipment (1.1 per cent). This pattern of domestic production contrasts with the domestic market, where industrial products (including office automation equipment) made up 31 per cent of sales, and exports, where their share was only 7 per cent.

The available evidence is therefore hardly overwhelming. Moreover, it provides little insight into the type of automation involved. Were the numerically controlled machine tool sales to LDCs part of intra-sphere automation systems, or were they confined to particular sectors? In the absence of determinate information it is not possible to draw any conclusions. But perhaps there is evidence of comparative advantage reversal in final markets?

As in the case of the diffusion of automation technologies, evidence that the *pattern of trade in manufactures* is being affected by the diffusion of automation technologies in developed countries is also limited, although there is no shortage of predicitions. It is interesting that the evidence which has emerged relates to the core, electronic-components industry itself and here a number of observers (for example Rada, 1982a and 1982b, and Ernst, 1982) claim to see the first signs of trade reversal. There two major reasons for this. First is the problem of market entry which has led to most of the major producers establishing plants close to the market, generally in Europe or America. As Rada (1982b) observes, it is no accident that in the 1979–82 period, 'The largest investments in integrated circuits (outside the companies' own countries) have taken place in Scotland' (p13). Second, there have been major changes in technology, almost all of which involve greater capital intensity and automation. Interestingly a primary factor conditioning these changes relates to the quest for greater quality. In the late 1970s Japanese penetration of the US semiconductor memory market reflected in large part their better quality. This resulted from their greater use of automated technologies in assembling the components. Whereas 37 per cent of the US firms' output involved some (labour-intensive) processing in low-wage economies, the equivalent figure for Japan was only 3 per cent. When Rada interviewed US firms in the early 1980s, he found few who considered it likely that they would establish new plants in the Third World. This, and other factors led Rada to conclude:

> Whilst smaller companies and new entrants might turn to subcontracting assembling operations off-shore [ie in low wage LDCs], as well as on-shore, rather than overstretch their managerial and financial resources with wholly-owned operations, the fact is that the larger the company and the more vertically integrated (which, as we have argued, appears to be the trend) the less the use of off-shore plants in terms of numbers as well as size . . . . As fixed costs became larger as a proportion of total cost and the value-added in semiconductor manufacture operations (as opposed to assembly) grows, we shall see an acceleration of the rationalisation in the present distribution of plants. In addition, as quality becomes crucial for competitiveness, assembly will change substantially with automation and the use of clean rooms. At this point fully integrated manufacturing plants [ie intra-sphere automation] will become more economical than the fragmentation of production (Rada, 1982b, pp12–13).

Beyond these events emerging in the semiconductor component industry itself, there is little sign that the introduction of automation technologies in developed countries has yet resulted in trade reversal. But the portents are there. Hoffman and Rush (1983) conclude that even in the garments industry – a sector which has been thought unsuitable for automation due to the limp nature of the material being assembled – the prospects for trade reversal are high. In the television industry, automated-insertion devices have clearly made it economically feasible for production to occur in Europe and North America, despite high wages (Sciberras, 1979). And in many other sectors, particularly the automobile industry, the introduction of automation technologies has almost

Table 9.3 DC imports of manufactures from LDCs in relation to design and draughting intensity

	Value $ million 1970	1978	*Growth* 1978/1970	*Rankings (N=15)* Value (1978)	Growth	Draughting intensity	Design intensity
Major traditional manufactures							
Semi-finished textiles	1 815	9 610	5.3	1	13	11	11
Leather	183	950	5.2	9	14	13	12
Clothing	1,181	9,502	8.1	2	10	12	14
Shoes	151	2,033	13.5	7	6	14	13
Major higher-technology manufactures							
Chemicals	588	2,282	3.9	5	15	9	6
Metals and metal products	319	2,223	7	6	12	10	9
Machinery except electrical and business	81	1 136	14	8	5	4	3
(Farm machinery)	2	29	14.5	15	4	7	7
Electrical machinery	372	4 463	12	3	7	1	1
Business machines	81	600	7.4	12	11	2	5
Scientific instruments	24	359	15	13	3	3	4
Motor vehicles	23	603	26.2	11	2	8	10
Aircraft	18	737	40.9	10	1	6	2
Shipbuilding	40	355	8.9	14	9	5	8
Consumer electronics	214	2,391	11.2	4	8	*	*
Total other manufactures	401	2,922	7.3				
Total major traditional manufactures	3,330	22,095	6.6				
Total major higher technology manufactures	1,762	15,178	8.6				
Total manufactures	5,493	40,195	7.3				

* Datedness of 1960 data does not allow for meaningful figures. In general design is high in this sector and so is draughting.

Source: Kaplinsky, 1982b

certainly been instrumental in preventing the migration of parts of the industry to low-wage economies. Similarly, the largest domestic appliance manufacturer in the world, situated in the USA, recently appraised the possibility of locating in the Third World in the context of emerging automation technologies. It concluded that whereas such a move might have been justified in the mid-1970s (and indeed had been undertaken by the competitor General Electric, amongst others), the availability of new automation technology in the 1980s made it unnecessary and indeed undesirable.

What is interesting is to attempt to project those sectors in which automation technologies are most likely to diffuse, and to examine this in relation to DC-LDC trade. There are few guidelines to use in this exercise and indeed few attempts have been made to do so. However, let us consider briefly the CAD sector. Here we can assume that the equipment diffuses most rapidly in those sectors which are design- and draughting-intensive, measured by the proportion of these skills in the workforce. Relating this to the pattern of trade in the 1970s it is clear that if the CAD sector is anything to go by, then automation technologies are most likely to diffuse first to precisely those higher-technology sectors in which LDC manufactured exports grew most rapidly in the 1970s and in which they plan to specialize in the 1980s (Table 9.3).

A 'first world technology'?

One of the major objectives of this book has been to clarify the evolution of automation technology. We observed that whereas previous rounds of automation have concerned factors such as the motive power involved (for example, steam, electricity or hydro-carbons) and the components of automation (control, transmission and forming) the significane of contemporary developments is to be found in the transition to systems-gains. Thus, in general the 1970s saw the maturation of electronics-based intra-sphere automation, especially in design and manufacture, which first showed the real potential for these systems-gains. The coming decades are likely to see the emergence of full, inter-sphere automation, leading to the 'factory of the future'. The key to these systems-gains is to be found in the linking together of individual, digital-logic, electronics control devices incorporated in separate activities. It is immediately clear that these systems-gains – especially those involving inter-sphere automation – are only open to those firms making wide use of electronics technologies. Merely introducing intra-activity automation – for example, numerically controlled machine tools – limits the potential gains which can be realized, and hence machinery manufactures in developed countries are evermore conscious to provide systems to their customers.

However, in LDCs, this pattern is not evident. Almost always indigenous technical change is concerned with overcoming industrial constraints and bottlenecks. An interesting example can be drawn from the sugar industry in which Indian technicians have recently made significant technological improvements (Kaplinsky, 1983a). Their major concern has been with the characteristics of their cane-crushing and juice-boiling activities. These have been improved, in unrelated ways, to increase the yield of sugar from cane, and to allow operation in wet weather. So successful have these advances been that a major alternative technology has been opened up for other low-wage, cane-sugar economies. However, the alternative developed-country technology, which boils the cane-juice under vacuum rather than atmospheric conditions, went through its intra-activity technological improvements in the nineteenth century. The concern of these machinery suppliers now lies in introducing electronic process controls (that is intra-sphere and inter-sphere automation) which reduce labour requirements and improve fuel-efficiency.

There are two linked points which emerge from this analysis. First, developed country machinery suppliers are increasingly concerned with capturing systems-gains, a possibility which has been opened up by the widespread diffusion of intra-activity, electronics-based automation technologies. And second, precisely because of this, developed country technology is increasingly irrelevant to operating conditions in LDCs. Thus, one of the major implications of the diffusion of contemporary automation technologies is likely to be a change in the nature and direction of global trade in manufacturers. Instead of technology flows going from North to South and being reflected in matching LDC exports of primary products, we are likely to see an increasing trend in intra-Third World trade in technology, particularly that which reflects intra-activity automation. Indeed, the past two decades have already reflected this trend, since the share of low-income LDCs merchandise exports going to other LDCs rose from 27 per cent in 1960 to 40 per cent in 1980; for middle-income, oil-importing economies, the rise was from 23 per cent to 34 per cent over the same period (IBRD, 1982). More recent and detailed evidence for individual economies such as Brazil and Argentina suggest that their technology and technology-intensive manufactured exports tend to go to other LDCs, whilst their traditional manufactures continue to go to industrialized economies such as the USA.

Concluding remarks on the impact of automation on export-oriented growth strategies

We have examined the likely impact of the diffusion of automation technology on export-oriented industrialization strategies, and have concluded that in addition to non-technology related factors (such as debt, militarism, the slowdown in world trade-growth and the 'fallacy of composition') there were indeed a number of technology-related factors which give rise to pessimism concerning such industrial strategies in the 1980s. Three factors were considered: namely the link between technology, economic crisis and market entry; comparative advantage reversal; and the emergence of a distinct 'First World Technology'. However, it is essential not to give too much autonomy to the concept of technology for it is not so much the evolution of the technology itself which gives rise to changes in comparative advantage, but the changing nature of social relations which determine the precise nature of technological progress, and hence the balance of comparative advantage in production. Let us consider this in a little more detail, although this subject is treated in considerably greater detail in the concluding chapter.

In the first place, the microelectronics technology which lies at the core of the new automation technologies did not so much rise out of an ordained technological evolution, but was continuously nurtured by the demands of the military sector. The first computers were built in the US and the UK for the military in the second world war; miniturization in the late 1950s was prompted by the Soviet launch of the first satellite; numerically controlled machine tool development in the 1950s was almost entirely due to the US Air Force; graphics software arose out of the missile-early-warning system; and so on. This process continues today, not only in the West but also in the Third World. The first Latin American Conference on the Impact of Microelectronics concluded that all of the first generation numerically controlled technologies in manufacturing were introduced to fill the needs of the military sector. As global arms sales become increasingly valuable and sophisticated, there is no doubt that this will give a spur to the diffusion of automation technologies in LDCs, not only in industry, but in relation to weapon systems in the armed forces as well.

Second, since the history of export-led growth has been so intimately bound up with the locational decisions of transnational corporations, it is instructive to assess their likely attitude to global location in the 1980s. Here it is important to conceptualize the relationship of firm-behaviour and technology in a meaningful way, for the interaction is of necessity a dialectical one. Thus, whilst decisions are made by management on the basis of the technological alternatives which are in fact available, at the same time these decisions also determine the technological alternatives which will be made available in the future. This is a subject we shall consider in greater detail in the final chapter. Here it is as important to ask ourselves whether the emergence of automation technologies is determined by the imperative for firms to adjust to a changing economic environment, as it is to ask how firms' location decisions are affected by technological developments.

In this regard it is evident that to a large extent the emergence of automation technologies reflects decisions forced upon transnational corporations by social and political developments. As we saw, the existence of increasing levels of unemployment is giving rise to irresistible tides of protectionism in all the advanced countries. What this protectionism has done is not so much to limit competition in the OECD, but to change its source, from low-wage economies (ie LDCs) and distant, high-wage economies (ie Japan), to proximate, high-wage competitors (ie in Europe or N. America). Thus, competition remains, but protection gives the opportunity to restructure and to introduce new technology. Hence for example, the Multi-Fibres Agreement (which regulates LDC exports of garments to the OECD) and the restrictions of motor-vehicle imports, both of which have given transnational corporations 'space' to develop automation technologies to wardoff future competition. We also observed a similar process in the electronics and television industries, where protectionist pressures have forced producers to locate near their final markets: this was only feasible if new, automation technologies could be developed and introduced. However, a contrasting factor affecting the locational decisions of transnational corporations concerns the nature of work-practices which are involved in using the new technology. In many cases the transition to systems-based technologies involves changes in the nature of work and a reduction in the skill-component in work (Chapter 8). As such there is likely to be some resistance from labour in developed countries, a 'problem' which is less likely to occur in new, greenfield sites in LDCs. Where this occurs, therefore, *and where politically feasible*, transnational corporations may prefer to locate in the Third World.

In addition to the forces of militarism and protectionism which condition both the nature and direction of technological progress being pursued by developed-country firms and transnational corporations, we must also consider the momentum of Third World governments and capitalist classes. They, too, often have an interest in pursuing export-led growth strategies. Indeed, contrary to popular myth, the governments of most successful Third World exporters have been highly interventionist, not only offering substantial financial incentives to exporting, but also coordinating investment decisions and pursuing policies designed to strengthen their electronics industries. Hence, even if market forces in themselves will not lead to the diffusion of automation technologies in the Third World, it is a reasonably safe assumption that many LDC states will put substantial resources into the acquisition of these technologies. The Brazilian government, for example, is currently pursuing an electronics-sector policy involving the expenditure of around $600m over five years. So that although market entry will remain a problem, and although the utilization of intra-activity automation technologies in the Third World is unlikely to allow the capturing of systems-gains, there will nevertheless continue to be substantial investments in automation technologies by a number of newly industrializing economies, thereby undermining to some extent the likely trend towards comparative advantage reversal.

In this context it is instructive to consider whether there are any inherent obstacles to users in LDCs taking advantage of the gains derived from these automation technologies. There are a number of points which can be made here. First, it is important to distinguish between various types and levels of skill. With regard to *management skills* it is clear that the new emphasis on systems-gains makes these capabilities of crucial importance. This concerns not only the extent to which management perceives the potential for systems-gains, but also their power in executing the desired changes in the face of opposition from suppliers, labour and the managerial class itself. For example, the new Chairman of General Electric,who, as we saw in Chapter 1, is trying to force the corporation to adopt new automation technologies, has insisted that all senior management attend re-educating engineering courses which illustrate the significance of systems-gains facilitated by new automation technologies. At the level of *operator* skills the evidence available suggests that the skill-threshold is lowered (Jacobsson, 1982; Kaplinsky, 1982b), making the technology relatively simple to use. With regard to *back-up* skills, particular abilities are required to repair electronics-based technologies. But offsetting this is the tendency for manufacturers to introduce systems which are self-diagnosing (those which identify the malfunctioning component allowing for simple replacement) and 'self-healing' (with redundant, additional systems which automatically come into operation when something fails).

Second, there are some signs that the new automation technologies are inherently more difficult to assimilate. In part this is because the relative absence of working parts makes 'reverse engineering' (that is, learning-by-undoing), a strategy used to great effect by the Japanese, more difficult. But it also reflects the greater difficulty in unpackaging technology. This is a phenomenon in which the technology-importer in LDCs attempts to unbundle the technological system and buy in individual items from different suppliers. These policies were very prominent in the late 1960s and early 1970s and in fact in some countries seem to have had the effect of speeding up the transfer of technology. However, as the new automation technology increasingly moves towards systems, then the potential for unpackaging – even in developed economies – is likely to be substantially diminished.

Finally, there is the question of maximizing links between technology suppliers and users, and between different users. The experience in many sectors which involve software-intensive automation technologies (for example, see Kaplinsky, 1982b on CAD) is that at least in the early years of the industry, it is important to have close links with the technology suppliers and other users who experience similar problems. Hence, there has been a tendency for 'user groups' to mushroom, exchanging both software and experience in dealing with the technology. Since few – if any – of the relevant suppliers exist in LDCs, it is questionable whether users in LDCs will be able to make maximum, effective use of the new automation technologies. Nevertheless, there is little evidence that fundamental obstacles exist to limit the use of the new automation technologies in the Third World. Whilst there are reasons to believe that particularly in the short to medium run, LDC firms will make suboptimal use of the new technology – and especially in the reaping of systems-gains – their capability will inevitably improve over time. Rather the major factor linking the new automation technologies to the frustration of export-led growth strategies, arises from the capability which the new technology gives the transnational corporations to adjust, at the least cost, to the changing global trading environment.

AUTOMATION TECHNOLOGIES AND BASIC NEEDS IN THE THIRD WORLD

In assessing the factors underlying economic 'development' (which in itself is a value-loaded concept) some observers have concluded that the exports of manufactures has played a strategic role in transforming the economic structures of developing economies. Indeed Westphal et al (1981) believe that South Korea's experience in expanding exports of manufactures – which as we saw was so remarkably successful – played a key role in diffusing industrial competence to other sectors. But for most developing countries – and even possibly for South Korea – what happens in foreign markets has little practical relevance for the basic needs of most of the population. They, as we saw in Tale 9.1, suffer from the most basic problems, being denied access to adequate food, water and shelter. What possible relevance do the new automation technologies have to meet these basic needs?

In theory, of course, the potential is enormous. The technology offers the opportunity not only to meet the basic needs of the world's population, but also to satisfy a host of other less urgent consumer needs. But the problem for the mass of the world's population is twofold: first, the technology is still immature, and, second, it tends to be owned 'appropriated' by corporations who are resident in the advanced countries and who are engaged in meeting the less urgent 'needs' of a minority of high-income consumers, largely based in the industrial economies. We shall consider these problems further in the following concluding chapter in which we will attempt an overall assessment of the new automation technologies. Here we briefly confine ourselved to the specific problems of the masses of the Third World, comprising between a half and two-thirds of the world's population.

For these people the potential which the new technology offers to produce consumer goods such as video and tape-recorders, motor cars and computer games is to all intents and purposes irrelevant, although many of them may either own or aspire to own (electronic) radios and watches. For them the primary importance of the new technology is therefore likely to be as a producer, rather than a consumer good. In this respect it is possible, even probable, that the industrial working class will be involved in utilizing the new automation technologies. However, for the rural masses, who generally comprise the majority of the population in LDCs, such exposure is less likely. Indeed in many countries, technology is so unevenly diffused that the rural population frequently does not even have access to the more basic items of agricultural technology such as metal ploughs, let alone to the more sophisticated automation technologies under discussion. The most likely impact on their lives is likely to be in an indirect form, as the new automation technologies, by reducing costs of production in advanced economies or in the industrial sectors of their own economies, lower the cost of agricultural capital goods. As the technology diffuses through to the 'service' industries, it is possible that microcomputers will be used in rural health centres to store case histories and as a diagnostic aid; or that they will be used to predict weather patterns. Another possible use may be in the utilities area, providing energy to rural enterprises; for example, recent developments in micro-hydro-electricity generators make extensive use of electronic switching devices to displace surplus energy.

There are, of course, more fanciful suggestions of ways in which the new technology can be utilized to meet the basic needs of the mass of the population in LDCs. One of these, promoted by a United Nations organization as a response to meeting basic needs, involves the installation of solar-powered video tape-recorders in Indian villages! But this is (perhaps thankfully) an unlikely eventuality and suffers, as do other suggestions of its ilk, from a naive view of the factors conditioning technical progress. These draw our attention back to the 'centre' of the world economy, to the environment in which the new automation technologies are being developed and diffused. It is an environment in which little attention is being paid to the needs of the mass of the world's population; for example, it is striking that of over 1,500 public conferences on information technology in the UK in 1982, only one (of six hours duration) addressed itself to the specific impact on the Third World. Unless we understand what drives this technical progress, we will be unable to appreciate why scarcity and need are likely to coexist with the unutilized power of a potentially liberating technology. It is to these issues which we now turn.

CHAPTER TEN

Technology or Society?

In attempting to gain a perspective on the new, electronics-based automation technology, we have traversed a great deal of ground. The analysis began by examining the link between the economic crisis now besetting the world and the diffusion of new automation technologies. By enabling the displacement of labour and hence lowering effective demand, there can be little doubt that the widespread use of these new technologies exacerbates the extent of economic decline. However, the primary causal link between crisis and automation was considered to be of a very different nature in which the descent into recession and depression occurred for autonomous reasons. As it became more severe, the supercompetitive environment forced firms into a much greater degree of 'efficiency', and the route to this 'efficiency' was to be found in the utilization of the rapidly maturing electronics technology in downstream manufacturing to improve both products and process. Thus the more conventional view that automation technology (by displacing labour) causes crisis should be turned on its head: instead the development and diffusion of the technology should be seen as a *response* to economic crisis.

Amongst the competing definitions of 'automation' discussed in Chapter 2, we have chosen to use it in its most general sense, that is as 'a synonym for advanced mechanization' (Einzig, 1957, p2). Given this wide perspective it is obvious that automation has been occurring for a considerable period of time. In the period before the nineteenth century, technological progress was concentrated on improving the 'transforming' process; then around the late nineteenth century, automation of handling became the focus for progress, supplanted in importance after the mid 1930s by the automation of the coordination function. The current wave of automation is, as we saw in Chapters 3–6, concerned with the reaping of systems-gains by linking together the various, disparate activities – including transformation, handling and control – throughout the enterprise. The primary, but not the only, facilitating technology in the current wave is electronics, whereby digital-logic control systems in a variety of separate activities are linked relatively easily, giving rise to synergistic systems-gains. For firms who make successful use of these new automation technologies, the gains are substantial; but what of less-successful firms (in both the developed and less-developed world), or the workers utilizing the new technology, or those displaced by it? And what of the impact of these myriad innovations on the economic systems in which they are introduced?

These and other similar questions were considered in Chapters 7, 8 and 9, when we discussed the impact of the new technology. We observed amongst other things, that its diffusion seemed to be associated with an increase in the concentration of ownership (although perhaps not necessarily in the concentration of production) and a growing dominance by international firms. With regard to labour, despite the increasing skills required to produce the technology and to manage its introduction, the consequence for the rest of the labour force has been less rosy. Most significantly it deskills work, particularly in the craft-intensive batch industries which comprise the bulk of the engineering industries output, and in office and 'thinking' work. Moreover, at an aggregate level there is now little doubt that despite public pronouncements to the contrary (as for example by the British Prime Minister's 'Think Tank' in 1979) the new technology does displace labour. Its diffusion will almost certainly be associated with substantially higher levels of unemployment, even in successfully innovating economies like Japan. All of these negative factors are of course exacerbated for the Third World. For their populations the new technology is at best irrelevant, at worst a threat to export-led growth strategies.

But how can it be, that a technology which promises so much, can offer so little to the mass of the world's population? After all surely any technology which reduces the need for human labour must in principle be favourable? For if our goal is to work for fulfilment rather than out of necessity, and if we are to be able to develop a wide range of skills and interests to satisfy ourselves as human beings, then clearly the advance of automation technology must be desirable. So how do we resolve this seeming paradox that the very technology which offers the unique potential for opening out our lives does so at the expense of most of the population's welfare? In order to proceed with the discussion we need to consider first what is meant by 'technology', and how it relates to social organizations.

TECHNOLOGY, TECHNIQUE AND SOCIAL RELATIONS

We have deliberately used the word 'technology' in a very loose way so far, interchanging not merely between the singular and plural forms, but also implicitly imbuing it with a wide range of meanings. It is now time to be more precise in the discussion. The key distinction lies between the concepts of 'technology' and 'technique' (Emmanuel, 1983). The former should be seen in its widest sense as a set of knowledge in a particular area which provides the principles under which applications can be developed. By contrast, 'techniques' are specific, concrete ways of combining people, machines and inputs to produce particular products or services. Stewart, quoting Merrill, refers to technology as the 'skills, knowledge and proceedures for making, using and doing useful things' (p1) and then distinguishes it from technique which:

> is associated with a set of characteristics. These characteristics include the nature of the product. The resource use – of machinery, skilled and unskilled, manpower, management, materials and energy outputs – the scale of production, the complementary products and services involved, etc (p2).

Thus often in this book we have really been using the wrong nomenclature, talking of 'technology' and 'technologies', whilst we should have been referring to 'technique' and 'techniques'. For example, microelectronics really is a technology, since in its broadest sense it refers to solid-state, binary-logic devices in general. By contrast, a numerically controlled machine tool, or a computer-aided design system, whilst both making extensive use of electronics technology, are actually specific techniques developed to perform particular tasks, and embodying particular types of work processes.

This is an important distinction since it goes a long way in helping us to resolve the paradox noted earlier, which was that the very technology which offers the potential to liberate human beings has in fact had the consequence of making life considerably more uncomfortable for the mass of the population. Thus it is not so much the technology which is problematical, but the particular bundle of techniques which have been developed. The issue at stake then is whether this particular set of techniques was an inevitable and unique development of the potential offered by the broad family of electronics-technologies, or whether alternative, viable techniques could have been developed. If the latter is the case, then we need to form a view of the conditions under which the development would have been possible. What factors then are important in shaping the bundle of techniques which are generated from any major set of technology?

There are a variety of competing explanations of the factors determining the nature of technology and techniques being developed, and why they are introduced. These can be grouped into two major sets. The first reflects a view of technological autonomy, in which technological progress follows a momentum of its own – this can be characterized as a form of 'Technological Darwinism'. The second, contrasting perspective seeks to root technology and techniques within the pattern of social relations. Here it is argued that different modes of production (for example capitalism or socialism), different sets of gender relations and the differential militarization of societies all affect the nature and diffusion of technological progress. Identifying which of these two perspectives represents the real world offers a number of rewards, the key one for our interest being whether the automation techniques we have been examining are in some sense an inevitable development, or whether they reflect the societies in which they are generated. If the latter is the situation, our concern must lie with the way in which our society is organized rather than with the technology itself.

'Technological Darwinism'

This phrase is used as a catch-all category for those observers who either explicitly or implicitly see technological progress as in some sense being 'neutral' in the relationships between technology and society, responding to an internal logic of its own. Techniques are developed in the quest for greater efficiency and those which are unable to withstand the test of competition are discarded, much as animal species which are unable to fend for themselves, are condemned to destruction. Thus only the fittest, 'efficient' techniques will survive and prosper, and hence the characterization of this approach as one of Technological Darwinism. This perspective can be variously rationalized in the following terms.

There are a variety of approaches which implicitly or explicitly consider that the nature of technological progress stands outside social relations. Sometimes it is considered as a given datum whose origins require no explanation; this is an approach, as we saw in Chapter 8, which is deeply inbedded in economic theory. But a more sophisticated view is one which explicitly recognizes the internal momentum of scientific and technological developments. Broad technological developments as well as some particular techniques result from the constant search by scientists and engineers to circumvent technological challenges. Thus, in the contemporary period, engineers face the challenge of producing suitable sensors, or robots with a visual-recognition capability – these developments are direct responses to the technical difficulties associated with the introduction of the new automation technology. Another variant is the view that technological progress results from the necessity to overcome particular bottlenecks, generally because a long-used input or product becomes either unavailable or high priced. An oft-cited example is the development of beet-sugar technology by the French during the Napoleonic wars. But in general, the common theme to all these approaches is that the determination of technological progress is somehow *exogenous* to the nature of social organization, except in the narrow sense that scientific institutions tend to have particular sociological characteristics (such as a quest for complexity) which affects the nature of the technology produced.

Once developed, inventions are commercialized through a process of *innovation*, in which entrepreneurs (or enterprises) apply them in competitive situations. New products or processes are thus introduced in the quest for monopoly profits, but these profits are undermined as competitors either emulate or supersede them with more 'efficient' variants. Firms which are unable to replicate or improve these technological innovations (which include organizational structures) face a collapse of profits and are forced out of production. This approach, commonly associated with the works of Schumpeter, gives a key role to technological progress in the process of economic expansion; but implicit in it is a view of 'efficiency' which proximates to the characterization of Technological Darwinism described above.

There is also the view that the factor determining the nature of technological progress – that is the characteristics of the particular bundle of techniques developed from a technology – tends to be the relative prices of inputs and factors of production. In fact there is an extensive debate in economic theory as to whether it is the existing or the anticipated price and availability of these inputs and factors which is relevant, but for the purposes of this discussion there is little difference between the various approaches. Their common theme is one which ascribes almost sole causality to relative prices and factor and input substitutions. The nature of the work processes involved, or the organization of the production, is considered irrelevant.

The underlying theme of the Technological Darwinist approach is thus one in which the social context of innovation is not considered as having any bearing on the question of technological progress. Their techniques are either assumed to be available or said to be the consequence of exogenous factors or inherent technological dynamism. Once introduced, the survival and extension of a technique is a function of its 'efficiency'. What is meant by this term is seldom questioned, the implicit message being that it reflects the (long or short-term) profitability to the producers.

With this perspective it really does make sense to talk of the impact of technology on society. In concrete terms, numerically controlled machine tools are more 'efficient' and will therefore inevitably become dominant; they displace the deskill most of the labour force, and therefore the comparative advantage of Third World producers is undermined. Considered in this way, technology assumes the blame for the disturbing developments which the world is experiencing and which we considered in previous chapters. And the implied response for those affected adversely by their use offers little scope outside of two alternative decisions. Either accept the techniques, despite their harmful impact upon some parts of the community, or resist their implementation, insulate the economy (or firm) from competing enterprises using the new technology and accept a lower standard of living.

Technologies and techniques as social products

Counterposed against the perspective of Technological Darwinism are a series of approaches which recognize that both technologies and techniques are social products. They represent the end-result of human endeavour and, as such, their make-up reflects the manner in which human beings are organized. Although there is a common thread which runs through these various approaches, there is inevitably also a difference in view between them. We focus on three sets of discussions (that is with regard to the labour process, the impact of militarism and gender relations) to illustrate the view that technology and techniques can best be considered as social products rather than as the end-result of an autonomously determined path of technological progress.

Technology and the labour process. The *labour process* perspective arose directly out of the writings of Marx in the nineteenth century, but has recently begun to have a significant impact upon non-Marxian sociological and innovation theory. Its major contribution has been to remove the focus from technology as a thing-in-itself and to relate it directly to the social formation in which it is generated and innovated. Technology is, in this perspective, never neutral, but reflects the power relations inherent in the social system. Again we can caricature the discussion as follows.

The analysis of technology and social relations must begin with a focus on social relations. In this it is imperative to recognize that individuals do not interact as atomistic units, but together with other people in the pursuit of common interests. Progress and change in any system is driven by the systematic attempts of these groups – called 'classes' or 'fractions' of classes – to maintain or to obtain dominance and a perpetual struggle for power ensues. Power is obtained and reinforced by the ability to command and appropriate surplus, that is the balance of production which is not consumed by the workforce. Technology offers the potential for this surplus to be generated and appropriated. But in order for this to be actualized, specific techniques are required. A variety of alternative techniques are in principle available. But 'alternative' here has an additional, different meaning to that implicit in economic theory which characterizes choice as involving different combinations of factors or production. In this case the alternatives referred to include the nature of the final product and the work processes involved in production.

Thus, according to this argument, 'Capitalist labour processes' are designed to maximize the profit to the capitalist by enabling him to generate and appropriate surpluses. (This is the 'efficiency' implicit in the Technological Darwinist school). In so doing they necessarily involve a minimization of labour input, the deskilling of labour so as to cheapen its cost (the so-called Babbage principle discussed in Chapter 8), a hierarchical system of control and an increasing subdivision of labour. However, this only represents a particular form of 'efficiency' and reflects the capitalist mode of production in which the technology is being generated and innovated. When labour is no longer scarce – and for how many contemporary societies is this not the case? – then the minimization of the labour input may no longer be desirable; when labour is renumerated relatively equally, then the necessity to deskill tasks is removed; and when conflict is no longer endemic at the point of production (conflict at the point of production being at the heart of capitalist systems) then the need to control by hierarchy and the fragmentation of tasks is removed. Furthermore, other undesirable developments associated with the introduction of the new automation technologies – such as the increasing concentration of ownership and the neglect of Third World needs – are a function of capitalist social organization rather than the inevitable consequence of using automation technology.

Thus, for the labour process school, the emerging electronics-based automation techniques are merely one particular bundle derived from a wider potential offered by the new technology. Indeed the very existence of sophisticated electronics control devices reopens the possibility of enskilling the labour process and a number of attempts are currently being made (some of which are outside the Marxian fold – see Chapter 4) to redirect technology in this way. Therefore, crucially for our purposes, an approach such as this gives little weight to the view that the new automation techniques are having a negative impact on society; instead the focus is changed to the social context in which the new techniques are being generated and diffused.

The military imperative and technology. The hidden hand underlying many of the technological developments described in earlier chapters has been the key role played by the *military*. It is essential therefore to examine the role played by the '*Military Imperative*' in shaping the nature of technical progress. To begin with it is necessary to have some overview of the extent and growth of global military expenditure. Consider the following data compiled by the United Nations (UN, 1982):

- World military expenditure in 1980 (between $510b and $630b) was 15–20 times more than total official development aid to developing countries.
- Even in the 1960s, which was not only a period of 'detente' but also saw the end of the Vietnam War, total global military expenditure grew (in real terms) at 1.8 per cent pa.
- Some 50 million people are involved in the military sector, of whom only half are soldiers.
- In some sectors and countries the military presence is particularly pronounced. In 1977, over half of total US aircraft and 46 per cent of ship-building went to the military; the military accounted for 46 per cent of aerospace sales in France (1977), 70–80 per cent in Germany (1977) and 50 per cent in the UK (1980).

Most significantly for the purposes of this discussion, the UN report concluded that:

> It is in the field of science and technology, however, that the diversion of resources to military ends is greatest, Recent estimates indicate that up to 20–25 per cent of the world's scientific and research activities are directed towards military purposes. The research intensity of the average military product is some 20 times higher than that of an average non-military manufactured product (p204).

Clearly, therefore, a perspective on the rate and nature of technological growth must give great weight to the influence of the military imperative. This is particularly true of the new automation technologies which we have discussed in earlier chapters (especially Chapters 3 and 4). For example the first computers were developed to meet military needs in the Second World War; numerical control machine tools were specifically developed for the military (who paid not only for the R & D but also subsidized the first users); miniaturization of elec-tronic circuits was a specific military response to Soviet space successes in the 1950s; the basis of computer-aided design was laid with the early-warning missile system in the 1950s; the US Air Force currently has a large programme (costing over $100m) to encourage the use of automation techniques by Amer-ican industry, even in those firms not producing directly for military needs; the US Department of Defence is sponsoring US research in the area of Super-Very-Large-Scale-Integration, the next generation of integrated circuits; and so on. Therefore it is of relevance to try and set out the systematic ways in which the military imperative conditions the process of technological development. This can be crudely abbreviated in the following terms.

Militarism cannot be seen as a homogeneous concept and is made up of four major phenomena (Smith, 1983). These are high levels of military expenditure, the militarization of domestic social relations, the use of war and force in international relations, and the development of nuclear weapons. Each of these forms of militarism has its own specific historical roots and can exist in isolation, in tandem, or grouped together. And each has its own requirements for military technology and specific techniques (ie weapons).

The rise of the military sector over the years has been intimately bound up with the evolution of the modern centralized state. As MacKenzie (1983) points out:

> . . . the relationship between the creation of 'modern' armies and the creation of 'modern' states is far from accidental. Both, too, are closely interwoven with the creation of the 'nation', for the modern army is of course a national army, and the modern state predominantly a nation state . . . (p57).

As a consequence both the military and the centralized state are pervasive in most contemporary societies and hence play an important role in allocating productive and inventive resources. An alliance between elements of the state and the productive system gives rise to the military-industrial complex, which exists to coordinate and press the claims of those benefiting from military activities.

The military-industrial complex appears to exist across modes of production, a point recognized by Marxists (Jahn, 1975; MacKenzie, 1975, 1983; Georgiou, 1983; Murray, 1983; Smith, 1983) and non-Marxists alike. We focus on its role in the capitalist system, since it is in this mode that the new automation technologies have originated and matured. Here, in recent years, the military sector has had both positive and negative impacts upon economic growth. On the positive side it is considered that militarization counteracts a tendency to declining rates of profit by providing protected areas of profitability (via cost-plus pricing) and by providing automation technologies which reduce the cost of fixed capital. Moreover, by 'wasting resources' it counteracts a tendency towards underconsumption and at the same time creates the conditions under which the capitalist class maintains its dominance. On the negative side, military expenditure contributes to inflation and diverts resources from productive sectors (Pavitt, 1980).

It is possible to observe a close link between the capitalist form of social organization, the military-industrial complex it spawns and the emergence and diffusion of automation technologies. Production for the military sector occurs under very different conditions than that for others sectors, in that characteristically it is free from competition and only a single purchaser is involved. As a consequence there appears to be tendency for great inefficiency in production – by some accounts the rate of investment is only half that in comparable non-military sectors, and both machinery and the labour force are relatively old (MacKenzie, 1983). Since, as we saw in our discussion of numerically controlled machine tools in Chapter 4, process technology has an important impact upon product performance, there is consequently a tendency for the military authorities to intervene in their supplying industries to induce them to upgrade the quality of the technology they produce and use. Hence the intervention of the US Air Force over the years in developing and diffusing numerically controlled equipment, CAD technology and other automation equipment.

It is further argued that in addition to this link between the capitalist mode of production and the development of automation technology there is what Kaldor (1982) has termed the 'baroque technology' syndrome. By this is meant the failure of military systems to adapt to the technology of the future but rather to concentrate on the technologies of their past successes. Hence the traditional British concentration on warships and the American concern with aerial technology, both affecting the allocation of inventive resources and hence the direction of technical progress.

Finally, as the professionalization of armies has grown over the years, so the labour process involved in these military sectors has changed, with important consequences for the technologies it requires. Consider, for example, the difference between pre-industrial and modern armies:

> The form of force that characterises advanced capitalist societies is the industrial army. One feature of the army is the weapons system concept. The armed forces, with the possible exception of the infantry, are organized around the weapons system, which comprises the weapons platform (the ship, aircraft, tank, etc) the weapons, and the means of communication. [By contrast] The pre-industrial army [is] . . . typically . . . infantry – or cavalry – based . . . the weapon being still the instrument of the soldier . . .
>
> The resulting organisation is hierarchical, atomistic and dehumanising. It reflects the importance accorded to industrial products, particularly machines in [advanced capitalist] society as a whole. Furthermore, the weapons are themselves ranked and subdivided into a hierarchical military organisation, minimising the possibilities for individual or small group action (Albrecht and Kaldor, 1979, p10 and 12).

The consequence is an increasing tendency towards technological complexity in modern military systems, which incidentally makes them particularly susceptible to systems breakdown (Kaldor, 1982). For example, a squadron of F–11 Phantom aircraft requires an inventory of around 70,000 spares. Thus with a similar level of real expenditure the US Air Force was able to purchase 3,000 tactical aircraft a year in the 1980s and only 300 in the late 1970s.

The conclusion which can be drawn from this line of discussion is that the stronger the military imperative the greater the proportion of inventive resources going into the development of military technologies. These technologies tend to have particular features, associated with greater complexity, sophistication and bigness and as a consequence have created the opportunity and environment in which particular types of techniques – in this case facilitating automation – are developed. Moreover, the removal of arms-production from the competitive arena has induced the military bureaucracy to intervene in production to speed up the development and diffusion of automation technologies. The most notable, documented case are numerically controlled machine tools and computer-graphics, but the tendency can be traced through the full spectrum of emerging automation technologies.

Gender relations, militarism and technology. In recognizing the technology reflects social relations, we have hitherto illustrated the discussion by drawing on two sets of literature, namely that concerning the labour process and militarism. We observed that whilst the mode of production clearly has an important impact upon the development of technology, it is not deterministic, and other factors, such as the type and degree of militarization, also intervene. Yet militarism too is an incomplete explanation, partly because as Smith (1983) points out it has different forms which are often historically specific. But more problematically we are left with a feeling that at a causal level other factors underlie the phenomenom.

To explain the prevalence of militarism, various competing explanations have been proffered, including the natural aggressiveness of the human species and particular systems of belief. One relatively recent perspective is that which focuses on *gender relations* and argues that high levels of militarism (and hence the techniques which are developed by societies) are associated with particularly patriarchal systems. Easlea (1983), for example, argues that 'the nuclear arms race is in large part underwritten by masculine behaviour in the pursuit and application of scientific inquiry' (p5). Nowhere is this illustrated better than in the testing of the hydrogen bomb in which the successful blast was announced by the code, 'Its a boy'. (Presumably the code for a failed attempt would have been, 'Its a girl'.) In other areas of investigation, it is believed (for example Eriksen, 1963) that the culture of militarism is underwritten by authoritarian and partiarchal family structures. In 'art-studies' too this issue is now getting more attention. It is perhaps no accident that throughout the ages in the Western cultural arena, the image of 'peace' in the visual arts has almost always been feminine, as opposed to the representation of 'war' as a masculine attribute (Llewellyn, 1983). Finally, in development studies there is a burgeoning literature on the way in which particular techniques are produced and used to underwrite the dominance of male-interests in the home and in agriculture (E. Boserup, 1970, A. Whitehead, 1981).

Our purpose, here, is not to argue that automation technologies only reflect the pattern of gender relations in societies. Even if this simplistic view were basically sound, there is no doubt that the area is inadequately reasearched for this point to be made conclusively. As MacKenzie (1983) states in concluding his review of the literature on gender and militarism in the following terms:

> I am painfully aware that these remarks on gender scarcely scratch the surface of the topic. They are in a sense simply a plea for further work. The relations between partriarchy and capitalism have received a gread deal of theoretical attention in recent years. Those between patriarchy and the militarist state are deserving of no less (p60).

Rather what we have been concerned to show is that the degree of militarism in any one society is not predetermined, but is open to variation by factors which include gender relations. Once this link is drawn, it provides further evidence for the basic view we are proposing which is that technological progress is not determined by autonomous forces, but is shaped by the nature of social organization.

THE IMPACT OF THE NEW AUTOMATION TECHNOLOGIES ON SOCIETY

In the light of these conclusions we must challenge the relevance of the question commonly at the forefront of public discussion, namely, 'What is the impact of the new automation technologies on society?' For our central concern should lie with the society we live in, rather than the technology it produces. In reality there is nothing essentially wrong with the new technology – by providing the capability to process and communicate information very rapidly and at very low cost it offers enormous potential. At last we can see the prospect of global society producing enough to meet and transcend basic needs. Yet the particular techniques which are being developed, responding to the needs of the dominant actors in advanced capitalist and patriachal societies, seem to be associated with a reduction in welfare for the mass of the population, be it in developed or developing economies. To focus resistance on the techniques which this society produces, rather than on the nature of the society which produces these techniques is to mistake cause for effect. In this case social *re*-action – whether it be to 'rehumanize' technology or to resist or destroy technological innovations – will inevitably be misplaced.

Finally, we return to the theme with which we began this book. In Chapter 1 we argued that it is not the diffusion of automation technique which is causing the economic crisis currently plaguing the world economy, an economic crisis which sees growing unemployment, a decline in the real value of trade and growth in almost all economies, and which has already in some cases reached the proportions of the 1930s. Rather, it was the emergence of crisis in the advanced capitalist countries in the early 1970s, exacerbated by the sudden twist in oil prices, that gave rise to the super-competitive pressures which led to the development and diffusion of the new, electronics-based automation technologies.

Nevertheless whilst recognizing this direction of causality, it would clearly be mistaken not also to recognize the interaction between automation technologies and the economic crisis which the world economy is now experiencing. Although the basic cause of economic crisis is extraneous to the development of the new technologies, there is some form of feedback and their diffusion is undoubtedly exacerbating the situation. With each successive dimunition in the rate of growth (and in some cases the absolute size) of demand, markets become more competitive, and increasingly automated techniques are introduced. However, their very introduction, by reducing the requirement for labour and other inputs, only results in a further dimunition of demand, and a descent into deeper crisis.

Once again our attention must be focused on the nature of social organization. For even this problem of demand-management – surely a necessary prelude to economic recovery – is one which concerns the way in which we structure our society. Whether it by by stimulating demand, either in the First or the Third World, or a reduction in the average working week, coordinated social action is required. However, the political forces ascendanct in almost all the major economies, with their orientation towards 'markets', makes such coordinated action seem unlikely, at least in the short to medium run. And even then, the 'problem' which are emerging and were documented in Chapters 7, 8 and 9, go beyond tinkering with the level of aggregate employment. It must therefore be debatable whether without fundamental changes in the way our society is organized, we can ever come to terms with a more 'humane' and less crisis-laden pattern of technological progress.

Underlying this discussion is a frightening set of developments which warns against complacency. The last two decades have seen a steady expansion of military forces, not just in the aggregate resources going into their production, but more importantly in the exponentially increasing destructive power. We stand on the verge of annihilating all life, and the stakes are terrifyingly high. In this context it should not be forgotten that the Second World War arose out of the depths of the last long wave depression; the US economy for example only regained its pre-depression level as war-production expanded (Galbraith, 1959). Clearly it is inconceivable that we should allow events to take the same course again.

So whether our concern lies in the narrow territory of the nature of work, or in the broader question of the continued survival of life on earth, it is essential to come to terms with the link between the new technologies and social relations. Neo-Luddism – the response drawn by many whose jobs are being destroyed by the technology's diffusion – is not just futile, since international firms find it relatively easy to relocate production elsewhere, but also seriously mis-specifies the analysis. The problem lies not with the technology, but in a form of social organization which misuses its potential to produce frighteningly destructive weapons, inappropriate products and undesirable work processes. This stands as a direct indictment of contemporary capitalism. And whilst it would be foolish not also to consider the potential nature of 'socialist-technology' (stripping away the naivété of 'no conflict in the workplace perspectives') this is a separate task and one which can wait. For the moment we are confronted by the momentum of capitalist social relations and the techniques it develops and diffuses. They threaten our very existence, and the future looks bleak unless we strike out at the roots of their development, rather than focusing on the technological forms which are produced.

Bibliography

Adams, G., 1982, *The Politics of Defence Contracting: The Iron Triangle*, Transaction Books, London and New Brunswick.

Advisory Council for Applied Research and Development, 1979, *Joining and Assembly: The Impact of Robots and Automation*, HMSO, London.

Albrecht, U., and Kaldor, M. (eds), 1979, *The World Military Order*, Macmillan, London.

Allison, C., and Green, R. (eds) 1983, *Accelerated Development in Sub-Saharan Africa*, Bulletin of The Institute of Development Studies, vol 14, no 1., Brighton.

Amber, G. H. and Amber, P. S., 1964, *Anatomy of Automation*, Inglewood Cliffs, Prentice Hall, N. Jersey.

Anderson, K., and Baldwin, R. E., 1981, *The Political Market for Protection in Industrial Countries: Empirical Evidence*, World Bank Staff Working Paper no 492, International Bank for Reconstruction, Washington.

ANZAAS, 1979, *Automation and Unemployment: Papers presented at an ANZAAS Symposium*, The Law Book Co., Sidney.

Archer, J., and Lloyd, B., 1982, *Sex and Gender*, Penguin, London.

Association of Professional, Executive, Clerical and Computer Staff, 1979, *A Trade Union Strategy for the New Technology*', reprinted in T. Forester (ed), 1980.

Audits of Great Britain, 1980, *Home Audit Facts and Figures*, AGB Research, Ruislip, Middlesex.

Ayres R. U., Miller, S. M., 1981, 'Robotics, CAM and Industrial Productivity', *National Productivity Review*, vol 1, no 1, pp 452–60.

Ayres R. U., Miller, S. M., 1983, *Robotics: Applications and Social Implications*, Ballinger Publishing Co., Cambridge, Mass.

Babbage, C., 1832, *On the Economy of Machinery and Manufacture*, Charles Knight, Pall Mall East, London.

Bagrit, L., 1965, *The Age of Automation: BBC Reith Lectures 1964*, Weidenfeld and Nicolson, London.

Bain, J. S., 1956, *Barriers to new competition: Their character and consequences in manufacturing industries*, Harvard University Press, Cambridge, Mass.

Balassa, B., 1980, *The Process of Industrial Development and Alternative Development Strategies*, World Bank Staff Working Paper No. 438, International Bank for Reconstruction and Development, Washington.

Barr, K., 1979, 'Long waves: a selective, annotated bibliography', *Review* vol 11, no 4, Spring.

Barron, I., and Curnow, R., 1979, *The Future of Information Technology*, Frances Pinter, London.

Basset, P., 1979, 'Labour and The Microchip', *Financial Times*, 23 October.

Bednavik, K., 1965, *The Programmers: Elite of Automation*, MacDonald, London.

Bell, R. M., 1972, *Changing Technology and Manpower Requirements in the Engineering Industry*, Sussex University Press, London.

Berle, A., and Means, G. C., 1932, *The Modern Corporation and Private Property*, Macmillan, New York.

Bessant, J., 1980, *The Influence of Microelectronics Technology*, Technology Policy Unit, University of Aston, Birminghan.

Bessant, J., and Lamming, R., 1983, 'Some Management Implications of Advanced Manufacturing Technology', mimeo, Department of Business Studies, Brighton Polytechnic, Brighton.

Betts, P., 1981, 'What you now have is a unique company', *Financial Times*, 17 August.

Beveridge, W., 1944, *Full Employment in a Free Society*, Allen and Unwin, London.

Blair, J. M., 1972, *Economic Concentration*, Harcourt Brace, New York.

Blauner, P., 1964, *Alienation and Freedom: The Factory Worker and his Industry*, University of Chicago Press, Chicago and London.

Bloomfield, G., 1978, *The World Automative Industry*, David and Charles, London.

Boserup, E., 1970, *Woman's Role in Economic Development*, St Martin's Press, New York.

Braun, E., 1982, 'Electronics and industrial development', in R. Kaplinsky, 1982a, pp 5–13.

Braverman, H., 1974, *Labour and Monopoly Capital: The Degradation of Work in the Twentieth Century*, Monthly Review Press, New York.

Brett, E. A., 1983, *International Money and Capitalist Crisis: The anatomy of global disintegration*, Heinemann, London, and Westview, Boulder.

Bright, J. R., 1958, *Automation and Management*, Harvard University Press, Boston.

Brighton Labour Process Groups, 1977, 'The Capitalist Labour Process', *Capital and Class*, no 1, Spring, pp 3–26.

Brooke, M. Z., and Lee H. Remmers, 1970, *The Strategy of Multinational Enterprise: Organisation and Finance*, Longmans, London.

Burns, J. C., 1980, 'The Automated Office in the Microelectronics Revolution' in T. Forester (ed) 1980.

Business Week, 1980, 'GEs new input for "factories of the future"', 22 December, p 22.

Business Week, 1980, 'Attacking GE's grip on controls', 22 December, pp 66 B-1-66D-1.

Business Week, 1981, 'The Speedup in Automation', 3 August, pp 48–50.

Carnegie–Mellon, 1981, 'The Impact of Robotics on the Workforce and Workplace', Carnegie–Mellon University, Pittsburgh, Pa.

Caves, R. E., 1980, 'Industrial Organisation, Corporate Strategy and Structure', *Journal of Economic Literature*, vol XVIII, pp 64–92.

Cawkell, 1980, 'Forces Controlling the Paperless Revolution', in T. Forester, (ed), 1980.

Central Policy Review Staff, 1978, *Social and Employment Implications of Microelectronics*, HMSO, London.

Chandler, A. D., 1962, *Strategy and Structure: Chapters in the History of the American Industrial Enterprise*, MIT Press, Cambridge, Mass.

Chandler, A. D., 1977, *The Visible Hand: The Managerial Revolution in American Business*, Harvard University Press, Cambridge, Mass.

Chilton, C. H. (ed), 1960, *Cost Engineering in the Process Industries*, McGraw Hill, New York.

Clark, J., Freeman, C., and Soete, L., 1980. 'Long waves and technological developments in the 20th century', mimeo, Science Policy Research Unit, University of Sussex, Brighton.

Clark, J., and Cable, V., 1982, 'The Asian Electronics Industry looks to the Future', in R. Kaplinsky (ed), 1982, pp 24–34.

Commonwealth Secretariat, 1982, *Protectionism: threat to international order; the impact on developing countries. Report by a Group of Experts*, London.

Cook, M., 1972, 'Cut Emotion from Group Technology to Get at the Facts', *The Engineer*, 6 April, pp 29–30.

Cook, H. W., 1975, 'Computer Managed Parts Manufacture', *Scientific American*, vol 232, pp 22–9.

Cooley, M. J. E., 1977, 'Taylor in the Office', in R. N. Ottoway (ed).

Cooley, M., 1980, *Architect or Bee? The Human/Technology Relationship*, Hand and Brain Publications, Slough.

Coombs, R. W., 1982, *Automation and Long Wave Theories*, Phd Dissertation, University of Manchester.

Cooper, C. M., 1971, 'Science, Technology and Development', *Economic and Social Review*, vol 2, no 2, January, pp 165–189.

Counter Information Services, 1979, *The New Technology*, Anti-Report no 23, London.

Crouch, C., and Pizzone, A. (eds), 1978. *The Resurgence of Class Conflict in Western Europe since 1968, 2 vols*, Macmillan, London.

Datamation, 1981, 'Japanese Push Robotics', vol 27, no 7, pp 56–7.

Denison, E. F., 1962, The Sources of Economic Growth in the United States and the Alternative Before US, Committee for Economic Development, New York.

Dept. of Industry/National Engineering Laboratory 1977, *Automated Small Batch Production: Technical Study*, National Engineering Laboratory, Glasgow.

Dickson, D., 1974, *Alternative Technology and the Politics of Technological Change*, Fontana, London.

Diebold, J., 1964, *Beyond Automation: Managerial Problems of an Exploding Technology* McGraw Hill, New York.

Dolotta T. A., et al., 1976, *Data Processing in 1980–1985: A Study of Potential Limitations to Progress*, Wiley-Interscience, New York.

Easlea, B., 1983, *Fathering the Unthinkable: Masculinity, Scientists and the Nuclear Arms Race*, Penguin, London.

Einzig, P., 1957, *The Economic Consequences of Automation*, Secker and Warburg, London.

Electronic Industries Association of Korea, 1982, *Electronics Industry in Korea*, Seoul.

Elger, T., 1979. 'Valourisation and "Deskilling": A Critique of Braverman',*Capital and Class* no 7, Spring, pp 58–99.

Eltis, W. A, 1966, *Economic Growth: Analysis and Policy*, Hutchinson, London.

English, J. M. (ed), 1968, *Cost Effectiveness: The Economic Evaluation of Engineered Systems*, Wiley and Sons, New York.

Erikson, E., 1963, *Childhood and Society*, 2nd Edition, Norton, New York.

Ernst, D., 1982, *The Global Race in Microelectronics: Innovation and Corporate Strategies in a Period of Crisis*, Campus, Frankfurt.

European Trade Union Institute, 1980, *The European Economy 1980–85: An Indicative Full Employment Plan*, Brussels.

Forester, T. (ed), 1980, *The Microelectronics Revolution*, Basil Blackwell, Oxford.

Freeman, C., 1974, *The Economics of Industrial Innovation*, Penguin Books, London.

Freeman, C., 1979, 'The Kondratiev long waves, technical change and unemployment', in *Structural Determinants of Employment and Unemployment*, vol 2, OECD, Paris.

Freeman, C., and Curnow R., 1978, *Technical change and Employment – A Review of Post-War Research*, Paper prepared for Manpower Services Commission, Brighton, Science Policy Research Unit, University of Sussex.

Freeman, C., Clark J., and Soete, L., 1982, *Unemployment and Technical Innovation: A Study of Long Waves and Economic Development*, Frances Pinter, London.

Friedman, A., 1977, 'Responsible Autonomy versus Direct Control over the Labour Process', *Capital and Class*, no 1, Spring, pp 43–58.

Futures, 1981, Special Issue on Technical Innovation and Long Waves in World Economic Development, vol 13, no 4.

Galbraith, J. K., 1959, *The Great Crash*, Penguin, London.

Georgiou, G., 1983, 'The Political Economy of Military Expenditure', *Capital and Class*, no 19, Spring, pp 183–204.

Gershuny, J., 1979a, 'Technical Change and Sectoral Development', mimeo, Brighton, Science Policy Research Unit, University of Sussex.

Gershuny, J., 1979b, *Changing Value Patterns and Their Impact on Economic Structure*, Paris, OECD.

Gershuny, J., and Miles, I., 1983, *The New Service Economy*, Frances Pinter, London.

Gerwin, D., 1982, 'Do's and don'ts of computerised manufacturing', *Harvard Business Review*, March/April, pp 107–116.

Giedon S., 1948, *Mechanisation Takes Command: A Contribution to Anonymous History*, Oxford University Press, New York.

Godfrey, M., 1983, 'Export Orientation and Structural Adjustment in Sub-Saharan Africa'. C. Allison and R. Green (eds) 1983, pp 39–44.

Gorz, A. (ed), 1976, *The Division of Labour: The Labour Process and Class Struggle in Modern Capitalism*, Harvester Press, Brighton.

Griffiths, J., 1981, 'FIAT: new robots for engine assembly', *Financial Times*, 2 November, p 10.

Halevi, G., 1982, *The Role of Computers in Manufacturing Processes*, John Wiley and Sons, New York.

Haustein, H. D., and Maier, H., 1981, 'The Diffusion of Flexible Automation and Robots', *International Institute for Applied Systems Analysis*, Working Paper 152, Laxenburg.

Hayes, R. H., and Abernathy, W. J., 1980, 'Managing our way to economic decline', *Harvard Business Review*, July–August, pp 67–67.

Hazewindus, N., 1982, *The US Microelectronic Industry: Technical Changes, Industry Growth and Social Impact*, Pergamon Press, New York.

Helleiner G. K. (ed), 1982a, *För Good or Evil: Economic Theory and North-South Negotiations*, University of Toronto Press, Toronto.

Helleiner G. K., 1982b, 'International Trade Theory and Northern Protectionism Against Southern Manufacture', in G. K. Heillener, 1982a.

Henderson, J., and Cohen, R., 1979, 'Capital and The Work Ethnic', *Monthly Review*, vol 31, no 6, pp 11–26.

Hill, P., 1979, *Profit Shares and Rates of Return by Country*, OECD, Paris.

Hoffman, K., and Rush, R., 1982, 'Microelectronics and The Garment Industry: not yet a perfect fit', in R. Kaplinsky, 1982a.

Hoffman, K., and Rush, R., 1983, *Microelectronics and Clothing: The Impact of Technical Change on a Global Industry*, Geneva, ILO.

Hone, A., 1974, 'Multinational Corporations and Multinational Buying Groups: Their impact on the Growth of Asian Exports of Manufacturers – Myths and Realities', *World Development*, vol 2, no 2.

Husband, T., 1982, *Technology and Society: Outline Notes for Lectures*, mimeo, Imperial College of Science and Technology, London.

Ingersoll Engineers, 1980, *Industrial Robots*, Dept. of Industry/National Engineering Laboratories, Glasgow.

International Bank for Reconstruction and Development, 1981, *Accelerated Development in sub-Saharan Africa: An Agenda for Development*, Oxford University Press/IBRD, Washington.

International Bank for Reconstruction and Development, 1982, *World Development Report*, Oxford University Press/IBRD, Washington.

International Management, 1981, 'How General Electric plans to become the supermarket of automation', September, pp 37–40.

Jacobs, G., 1980, 'Design for improved value', *Engineering*, February, pp 179–182.

Jacobsson, S., 1981, 'Technical Change and Technology Policy: The Case of Numerically Controlled Lathes in Argentina', mimeo, Lund, Sweden.

Jacobsson, S., 'Electronics and the technology gap – the case of numerically controlled machine tools', in R. Kaplinsky, 1982a, pp 42–6.

Jahoda, M., 1980, 'Postscript on Social Change', in H. S. D. Cole et al, 1980, *Thinking about the Future: a critique of the limits of growth*, Chatto and Windus, London.

Jahn, E., 1975, 'The Role of the Armaments Complex in Soviet Society (Is there a Soviet Military Industrial Complex?)' *Journal of Peace Research*, vol XIII, no 3, pp 179–193.

Japan Foreign Press Centre, 1982, *Fiscal 1981 Survey on Labour Force by Occupation (Summary): Effect of Office Automation on Employment at Headquarters of Large Enterprises*, R-82-10, Tokyo.

Jenkins, C., and Sherman, B., 1979, *The Collapse of Work*, Eyre Methuen, London.

Jones, B., 1978, 'Distribution or Re-Distribution of Engineering Skills? The Case of Numerical Control', mimeo, Bristol.

Jones, D. T., 1980, 'The Metalworking Machine Tool Industry in Western Europe and Government Intervention', mimeo, Sussex European Research Centre, University of Sussex, Brighton.

Jones, D. T., 1983, 'Technology and Employment in the UK Automobile Industry', mimeo, Science Policy Research Unit, University of Sussex, Brighton.

Jones, T. (ed), 1980, *Microelectronics and Society*, Open University Press, Milton Keynes.

Kaldor, M., 1982, *The Baroque Arsenal*, Deutsch, London.

Kaplinsky, R., 1982a (ed), *Comparative Advantage in an Automating World*, Bulletin of the Institute of Development Studies, vol 13, no 2, Brighton.

Kaplinsky, R., 1982b, *Computer Aided Design: Electronics, Comparative Advantage and Development*, Frances Pinter, London.

Kaplinsky, R., 1983a, *Sugar Processing: The Development of a Third World Technology*, Intermediate Technology Publishers, London.

Kaplinsky, R., 1983b, 'Accumulation at the Periphery: A Special Case?' in R. Cohen (ed), *African Islands and Enclaves*, 1983, Sage, Beverley Hills and London.

Kay, G., 1979, *The Economic Theory of the Working Class*, Macmillan, London.

Klein S., 1982, *The Stanley Klein Newsletter on Computer Graphics*, vol 4, no 6.

Kline, M. B., and Lifson, M. W., 1968, in J. M. English (ed), 1968.

Kondratieff, N. D., 1935, 'The major economic cycle', *Voprosy Conjunktury*, vol 1, 1935, pp 28–79.

Kraft, P., 1977, *Programmes and Managers: The routinization of computer programmer in the U.S.*, Springer–Verlag, New York.

Lambert, R., 1983, 'Geared for a rail revival: General Electric in the U.S. goes for major investment', *Financial Times*, 14 April

Llewellyn, N., 1983, 'Art and Peace', mimeo, Brighton.

Lloyd, J., 1981, 'Consumer electronics job fall World-wide', *Financial Times*, 25 November.

R. Luckham and Campbell, H., 1982a, 'Militarisation Northern Perspectives', mimeo, Institute of Development Studies, University of Sussex, Brighton.

Luckham, R., 1982b, 'Militarisation and the New International Anarchy', mimeo, Institute of Development Studies, University of Sussex, Brighton.

Luckham, R., 1983, 'Armament Culture', mimeo, Institute of Development Studies, University of Sussex, Brighton.

Lund, R. T., Hall, L., and Horwich, E., 1977, *Integrated Computer Aided Manufacturing: Social and Economic Impacts*, Centry for Policy Alternatives, Massachusetts Institute of Technology, Cambridge, Mass.

Lund, R. T., Pinsky, D., and Schowalter, J., 1980, *Computer-aided Materials Processing: Report from Seminar*, Center for Policy Alternatives, Massachusetts Institute of Technology, Cambridge, Mass.

Mackintosh Consultants, 1977, *Yearbook of West European Electronics Data*, Mackintosh Publications, Luton.

Mackintosh, I. M., 'Micro: The Coming World War' *Microelectronics Journal*, vol 9, no 2, reprinted in T. Forester (ed), 1980, pp 83–109.

MacKenzie, D., 1983 'Militarism and Socialist Theory', *Capital and Class*, no 19, Spring, pp 3–73.

McLean, J. M., and Rush, H. J., 1978, 'The Impact of Microelectronics on the UK: A Suggested Classification and Illustrative Case Study', *SPRU Occasional Paper Series, no* 7, Science Policy Research Unit, University of Sussex.

Mair, L., 1966, *Primitive Government*, Penguin, London.

Marglin, S. A., 1976, 'What do bosses do? The origins and functions of hierarchy in capitalist production', in A. Gorz (ed), 1976.

Marx, K., 1976, *Capital Vol. 1* (Introduced by E. Mandel), Penguin, London.

Marsden, K., 1981, 'Creating the right environment for small firms', *Finance and Development*, vol 18, no 4, pp 33–36.

Maugham, S., 1976, 'What do bosses do? The Origin and Functions of Hierarchy in Capitalist Production', in A. Gorz (ed).

The Mechanisation of Work, 1982, *Scientific American*, vol 247, no 3.

Meredith, M., 1981, 'Hoover fears hard time', *Financial Times*, 15 September 1981.

Mowshowitz, A., 1976, *The Conquest of Will: Information Processing in Human Affairs*, Addison-Wesley, Reading, Mass.

Murray, F., 1983, 'The Decentralisation of Production – The Decline of the Mass-Collective Worker', *Capital and Class*, no 19, Spring, pp 74–99.

Myers, C. A. (ed), 1967, *The Impact of Computers in Management*, MIT Press, Cambridge, Mass.

National Computing Centre, 1979, *Impact of Microprocessors on British Business*, NCC Publications, Manchester.

National Economic Development Office, 1975, *Why Group Technology?*, HMSO, London.

National Machine Tool Builders Association, 1981, *Economic Handbook of the Machine-Tool Industry 1981–2*, McLean, Virginia.

Noble, D. F., 1979, 'Social Choice in Machine Design: The Case of Automatically Controlled Machine Tools', in A. Zimbalist (ed), 1979.

Northcutt, A., with Rogers, P., and Zeilinger, A., 1981a, *Microelectronics in Industry: Extent of Use*, Policy Studies Institute, London.

Northcutt, J., with Rogers, P., and Zeilinger, A., 1981b, *Microelectronics in Industry: Advantages and Problems*, Policy Studies Institute, London.

Northcutt, J., 1982, *Microelectronics in Industry: Whats Happening in Britain*, Policy Studies Institute, London.

Noyce, R., 1977, 'Microelectronics', *Scientific American*, September, reprinted in T. Forester (ed), 1980, pp 29–41.

OECD, 1979, *Interfuture: Research Report on The Future Development of Advanced Industrial Societies in Harmony with that of Developing Countries*, OECD, Paris.

OECD, 1980a, *Technical Change and Economic Policy: Science and Technology in the New Economic and Social Context*, OECD, Paris.

OECD, 1980b, *Information Activities, Electronics and Telecommunications Technologies – Impacts on Employment, Growth and Trade*, DSTI/ICCP/80.10, OECD, Paris.

Ottoway, R. N. (ed), 1977, *Humanising the Workplace: New Proposals and Perspectives*, Croom Helm, London.

Pavitt, K. (ed), 1981. *Technical Innovation and British Economic Performance*, Macmillan, London.

Pirenne, H., 1961, *Economic and Social History of Medieval Europe*, Kegan Paul and Routledge, London.

Plesch, P. A., 1978, *Developing Countries' Exports of Electronics and Electrical Engineering Products*, International Bank for Reconstruction and Development, Economics of Industry Division, Washington.

Pratten C. F., 1971, *Economies of Scale in Manufacturing*, Cambridge University Press, Cambridge.

Pratten, C. F., 1980, 'The Manufacture of Pins', *Journal of Economic Literature*, vol XVIII March, pp 93–96.

Pressman, R. S., and Williams, J. E., 1977, *Numerical Control and Computer-Aided Manufacturing*, J. Wiley and Sons, New York.

Rada, J., 1980, *The Impact of Micro-Electronics*, ILO, Geneva.

Rada, J., 1982a, *Structure and Behaviour of the Semiconductor Industry*, United Nations Center for Transnationals, N York

Rada, J., 1982b, 'Technology and the North-South division of labour', in R. Kaplinsky, 1982a, pp 5–13.

Rosenberg, N., and Frischtak, C. R., 1983, 'Technological Innovation and Long Waves', mimeo, Stanford University.

Rostow, W. W., 1978, *The World Economy: history and prospect*, Macmillan, London.

Rothwell, R., 1981, 'Structural Change and Manufacturing Employment',

Omega, vol 9, no 3, pp 229–245.

Rowthorn R., and Hymer, S., 1971, *International Big Business 1975–1967: a Study of Comparative Growth*, Occasional Paper, Department of Applied Economics, Cambridge University Press, Cambridge.

Sampson, G. P., 1980, 'Contemporary Protectionism and Exports of Developing Countries', *World Development*, February.

Samuelson P. A., 1981a, 'Final quarter appears due to fall far shy of third quarter', *Japan Economic Journal*, 10 March.

Samuelson, P. A., 1981b, '1980s and 1990s have potential for fairly good global economic growth', *Japan Economic Journal*, 17 March.

Schmitz, H., Technology and Employment Practices: Industrial Labour Processes in Developing Countries, Report prepared for Social Sciences Research Council, mimeo, Brighton.

Schumpeter, J. A., 1976, *Capitalism, Socialism and Democracy*, Allen and Unwin, London.

Sciberras, R., 1979, 'Technology Transfer to Developing Countries – Implications for Member Countries' Science and Technology Policy', *Television and Related Products Sector Final Report*, OECD, Paris.

Senker, P., Swords–Isherwood, N., 1980, *Microelectronics and the Engineering Industry: the need for skills*, Frances Pinter, London.

Senker, P., Huggett, C., Bell, R. M., and Sciberras, E., 1976, *Technological change, structural change and manpower in the UK toolmaking industry*, Engineering Industries Training Board, Research Paper 2, Watford.

Servan-Schreiber, J. J., 1968, *The American Challenge*, Hamish Hamilton, London.

Shaiken, H., 1980, *Computer Technology and The Relations of Power in the Workplace*, International Institute for Comparative Social Research, Discussion, Paper no 80–217, Berlin.

Smith, A., 1950, *The Wealth of Nations*, Everyman's Library, London.

Smith, R., 1983, 'Aspects of Militarism', *Capital and Class*, no 19, Spring, pp 17–32.

Soete, L., 1979, 'Technical Change, Import Penetration and UK Employment', Paper prepared for the Joint SSRC/IDS Conference on UK Employment Projections, Brighton, 24–25 May.

Stewart, F., 1978, *Technology and Underdevelopment*, Macmillan, London.

Stewart, F., 1982, 'Industrialisation, Technical Change and the International Division of Labour' in G. K. Helleiner, 1982a.

Stopford, J. H., and Wells, L. T., 1972, *Managing the Multinational Enterprise: Organisation of the Firm and Ownership of the Subsidiaries*, Basic Books, New York.

Taylor, F. W., 1903, *Shop Management*, New York, reprinted in F. W. Taylor, 1947, *Scientific Management*, Harper and Brothers, New York.

Taylor, F. W., 1911, *The Principles of Scientific Management*, New York, reprinted in F. W. Taylor, 1947, *Scientific Management*, Harper and Brothers, New York.

Thomas, H. A., 1969, *Automation for Management*, Gower Press, London.

Thomson, A. R. (ed), 1980, *Technology of Machine Tools: A Survey of the State of the Art by the Machine Tools*, Task Force, Lawrence Livermore Laboratory, Livermore, California.

Townsend, J., Henwood, F., Thomas, G., Pavitt, K., and Wyat, S., 1981, *Science and Technology Indicators for the UK: Innovations in Britain since 1945*, Science Policy Research Unit Occasional Paper Series, no 16.

Tzong-biau, Lin and Mark, V., 1980, *Trade Barriers and the Promotion of Hong Kong's Exports*, Chinese University Press, Hong Kong.

UK Central Computer Agency, 1978, *Word Processing by Computer: Report of a Pilot Project at the Dept. of Education and Science Darlington*, London.

UK Central Computer and Telecommunications Agency, 1980, *Stand-alone Word Processors: Report of Trials in UK Government Typing Pools 1979/80*, London.

United Nations, 1981, *Transnational Corporations and Transborder Data Flows: An Overview*, UN Economic and Social Council, Commission on Transnational Corporations, E/C. 10/87, New York.

United Nations Centre on Transnational Corporation, 1982, *Transnational Corporations in the International Auto Industry*, ST/CTC/38, New York.

United Nations Conference on Trade and Development, 1979, *The industrial policies of the developed market economy countries and the effect on the exports of manufactures and semi-manufactures from the developing countries*, TD/230/Supp. 1/Rev 1, United Nations, New York.

United Nations Conference on Trade and Development, 1980, *Trade in Manufactures of Developing Countries and Territories, 1977*, TD/B/C.2/187, United Nations, New York.

United Nations Conference on Trade and Development, 1981, *Trade and Development Report*, TD/B/863/Rev.1, United Nations, New York.

United Nations Conference on Trade and Development, 1982a. *Trade and Development Report*, ITDR/2., Geneva.

United Nations Conference on Trade and Development, 1982b, *Incentives for Industrial Exports*, TD/B/C.2/184, United Nations, New York.

United Nations Conference on Trade and Development, 1982c, *The Impact of electronics Technology on the Capital Goods and Industrial Machinery Sector: Implications for developing countries*, TDF/B/C.6/Ac.7/3, United Nations, Geneva.

UNDP/UNCTAD, 1979, *The Balance of Payments Adjustment Process in Developing Countries*, United Nations, New York.

US National Academy of Engineering, 1982, *The Competitive Status of the US Auto Industry*, National Academy Press, Washington.

Verein Deutscher Werkzeugmaschinen Fabriken, 1979, *International Statistics of Machine Tools*.

Verreydt, E., and Waelbroeck, J., 1980, *European Community Protection Against Manufactured Imports from Developing Countries: A Case Study in the Political Economy of Protection*, World Bank Staff Working Paper No 432, International Bank for Reconstruction and Development, Washington.

Walsh V., Moulton-Abbott, J., Senker, P., 1980, *New Technology, The Post*

Office and The Union of Post Office Workers, Union of Communication Workers, London.

Walton, R., 1975, 'Group Technology' *Work Study*, January, p 7–14

Westphal, L., Rhee, Y. W., and Pursell, G., 1981, *Korean Industrial Competence: Where it Came From*, World Bank Staff Working Paper no 469, International Bank for Reconstruction and Development, Washington.

Whisler, T. L., 1970, *The Impact of Computers in Organisations*, Praeger, New York.

Whitehead, A., 1981, 'A Conceptual Framework of the Analysis of the Effects of Technological Change on Rural Women', ILO *World Employment Programme Working Paper*, 2.22/WP.79, Geneva.

Wiener, N., 1948, *Cybernetics, or Control and Communications in the Animal and the Machine*, John Wiley and Sons, New York.

Wiener, N., 1950, *The Human Use of Human Beings*, Houghton Mifflin Co, Boston.

Wilkinson, B., 1981, *Technical Change and Work Organisation*, Phd Dissertation, University of Aston, Birmingham.

Williamson, J., 1983, 'The Economics of IMF Conditionality', in G. K. Helleiner (ed) 1982a.

Williamson, O. E., 1963, 'Managerial Discretion and Business Behaviour', *American Economic Review*, vol 53, no 5, pp 1032–57.

Wood, S. (ed), 1982, *The Degredation of Work*, Hutchinson, London.

Woodcock, C., 1983, 'Cloth makers weave a new plan for material benefits', *The Guardian*, 3 March.

Woodward, J., 1965, *Industrial Organisation: Theory and Practice*, Oxford University Press, London.

Yanemoto K., 1980, 'The Growing Market Size of Industrial Robot Industry', *Digest of Japanese Industry and Technology*, no 152 pp 25–31.

Zimbalist A., 1979, *Case Studies in the Labour Process*, Monthly Review Press, New York.

Author Index

Firm Index

Subject Index